Oil Palm's Water Use: New Measure

Oil Palm's Water Use: New Measure

Prof Marlon

First Printing, 2024

We've traveled a winding road, Horace,
full of unexpected turns

Table of Contents

CHAPTER 1 1

GENERAL INTRODUCTION ....

1.1 Tree and oil palm water use ....

1.2 Scaling of water use from tree or palm level to stand level ....

1.3 Spatial heterogeneity of tree and oil palm water use ....

1.4 Multi-level temporal dynamics ....

1.5 Outline and Objectives of the study ....

CHAPTER 2 7

Drone-based photogrammetry-derived crown metrics for predicting tree and oil palm water use ....

2.1 Introduction ....

2.2 Methods ....

2.2.1 Study area and sites ....

2.2.2 Sap flux measurements ....

2.2.3 Drone image acquisition and processing ....

2.2.4 Statistical analyses ....

2.3 Results ....

2.3.1 Plant water use ....

2.3.2 Drone-derived crown metrics and their relation with plant water use ....

2.3.3 Transpiration estimates and uncertainties ....

2.4 Discussion ....

2.5 Conclusions ....

CHAPTER 3 25

Airborne tree crown detection for predicting spatial heterogeneity of canopy transpiration in a tropical rainforest ....

3.1 Introduction ....

3.2 Materials and Methods

3.2.1 Study region and sites

3.2.2 Study plots and stand characteristics

3.2.3 Sap flux measurements

3.2.4 Remote sensing

3.2.4.1 Drone image acquisition

3.2.4.2 3D point cloud generation, individual tree crown detection and crown metrics

3.2.4.3 Automatic crown detection using AMS3D

3.2.5 Drone-based scaling, uncertainties and heterogeneity assessment of transpiration

3.3 Results

3.3.1 Tree water use vs. crown metrics

3.3.2 Individual tree crown segmentation

3.3.3 Canopy transpiration: scaling, uncertainties and spatial heterogeneity

3.4 Discussion

3.5 Conclusions

CHAPTER 4 43

Multi-level temporal variation of sap flux densities in oil palm

4.1 Introduction

4.2 Materials and methods

4.2.1 Study area

4.2.3 Sap flux measurements

4.2.3 Data analysis and statistical methods

4.3 Results and discussion

4.4 Conclusion

CHAPTER 5 58

SYNTHESIS AND OUTLOOK

5.1 Overview

5.2 Scaling variable and its associated uncertainties

5.3 Spatial heterogeneity of tree and oil palm water use

5.4 Radial flux and temporal variations of oil palm water use

5.5 Future scope ....

CHAPTER 1

GENERAL INTRODUCTION

1.1 Tree and oil palm water use

Transpiration (E_t) is a central flux in the terrestrial water cycle and is considered an important ecosystem service for atmosphere and hydrosphere regulation. Transpiration is strongly affected by land cover and land-use changes, which are currently very pronounced in tropical regions. In the lowland of Sumatra, large transformation of rainforest into monoculture oil palm plantations are widespread (FAO, 2016; Drescher et al. 2016) and have been associated with changes in the hydrological cycle; especially the transpiration (Merten et al. 2016; 2020). Also, higher transpiration rates from commercial oil palm were reported and may surpass the remaining forests (Meijide et al. 2018; Röll et al. 2019). Thus, it is important to understand the transpiration in oil palm monoculture and natural rainforest. Transpiration is commonly measured by sap flux techniques in the individual tree or palm (Granier 1985; Wullschleger, Meinzer, and Vertessy 1998). Sap flux method such as the heat dissipation method (Granier 1985) requires species-specific calibration to estimate the individual tree water use Lu et al. 2004). For oil palm, Niu et al. (2015) previously calibrated the specific constant to quantify oil palm water use and reported high transpiration rates from commercial oil palm plantations. Prior to our study, Röll et al., (2019) reported the trajectory of transpiration differences along with the land-use change where the transpiration rates decrease by 43 ± 11 % from forest to rubber monoculture but rebound with conversion to smallholder oil palm plantations and commercial oil palm plantation exceed high transpiration rates more than natural forest. Over the age gradients in oil palm plantations, the water use rates increase from 2 years old to 10 years old and then remained constant upto 22 years old (Röll et al., 2015). The tropical lowland area consists of undulating terrain and leads to upland and valley sites (Miettinen et al., 2014) and substantial heterogeneity of oil palm water use between riparian, partly flooded and upland sites were observed (Hardanto et al., 2017). In this book, we further extend the study of tree or palm water use in terms of scaling approach with the application of new drone technology, spatial heterogeneity of transpiration in tropical rainforest and temporal dynamics of palm water use for better understanding the transpiration patterns in near-natural and in human-modified ecosystems.

1.2 Scaling of water use from tree or palm level to stand level

Tree or palm water use is commonly measured by sap flux techniques, but the number of replicates is limited to cover a larger area (Granier, 1985; Wullschleger, Meinzer, & Vertessy, 1998). Later, individual tree water use is scaled up to stand transpiration through biometric variables. Thus, scaling is an essential step to estimate transpiration at stand level and with reduced uncertainties (Hatton and Wu 1995; Jarvis 1995; Moore et al. 2017). Tree diameter and tree density in the stand are the most common scaling variable since they are easy to assess through ground-based inventories (Wullschleger, Meinzer, & Vertessy, 1998). However, tree diameter has some limitations due to unexplained variability relating to water use and induced high uncertainties while scaling up the stand transpiration (Moore et al. 2017). In the case of oil palm, there may be a low intra-specific diameter variation but water use variations occur. Moreover, focus on transpiration studies in other monocot species such as bamboos have been increased in recent times (Mei et al. 2016). On the other hand, tree crown structure would be a potential candidate variable irrespective of monocot or dicot and also it is the interface of water exchanges with the atmosphere. It has been reported that the crown dimension scaled up well in a mature oak forest (Čermák 1989). Other studies also reported the very close relationship between crown structure and tree water use in *Taxodium distichum* forest and olive orchard (López-Bernal et al. 2010; Oren et al. 1999). In the premontane forest of Costa Rica, the transpiration was indirectly affected by crown exposure by influencing leaf wetness or dryness (Aparecido et al. 2016). However, it is difficult to measure crown dimensions through field inventory, particularly in dense and heterogenous tropical forest.

Modern applications of the drone in ecological studies have provided the feasibility to measure crown dimensions in the forest. A drone equipped with LiDAR (Lin, Hyyppä, and Jaakkola 2011; Wallace et al. 2012) or optical camera (Asner et al. 2002; Mlambo et al. 2017) has shown promising and new directions in forest inventories, particularly in assessment of crown and canopy structure (Barnes et al. 2017; Díaz-Varela et al. 2015; Thiel and Schmullius 2016). In recent times, a drone equipped with optical camera become popular due to low cost and high resolution imageris and capability to construct 3D point clouds through photogrammetry technique called Structure from Motion (SfM) (Dandois and Ellis 2010; Lowe 2004; Westoby et al. 2012) and further can compute tree crown structural metrics for relatively large area in a short time. Using this technique, several studies have computed various crown variables such as crown length (Kallimani, 2016), crown diameter (Lim et al. 2015; Panagiotidis et al. 2016), crown volume (Torres-Sánchez et al. 2015) or canopy cover (Khokthong et al., 2019) but few studies have reported in ecohydrological studies (Vivoni et al., 2014). The applicability of drone-based tree crown metrics as a scaling variable in stand transpiration estimates has not been explored yet. In our study, we derived several crown metrics based on drone photogrammetry techniques and tested against tree/palm water use to scale up the stand transpiration.

1.3 Spatial heterogeneity of tree and oil palm water use

Tropical rainforests comprise the dense and complex 3D structure, high tree species richness and diversity and covers heterogeneous site conditions (Whitmore 1998; Whitten and Damanik 2000). The overlapping and dense canopy structure may translate variable transpiration across the forest. Also, the quantification of canopy transpiration (E_t) in tropical forests may address the spatial heterogeneity of E_t and may increase the better understanding of the relationship between the structure and function of the tropical rainforest. Variability in site conditions also potentially reflects the spatial heterogeneity in rainforest E_t. Studies in boreal forest observed the differences in E_t by analysing the significance of site conditions for tree and stand E_t along the upland-to-wetland gradients (Angstmann et al. 2013; Loranty et al. 2008; Mackay et al. 2010). Thus, it is necessary to incorporate the heterogeneity due to topographic positions for landscape-level assessments. Such information about the heterogeneity are rare in tropical rainforest regions, but the control of water table on transpirations was only reported in few studies in northern Australia (McJannet et al. 2007) and in Hawaii (Santiago et al. 2000). In lowland of Sumatra, previous study reported substantial differences in E_t between upland and riparian sites for oil palm and rubber tree stands which is due to flooding (long-term and short-term) and topography (Hardanto et al. 2017); however no information is available for rainforest in the same region.

Predicing E_t across sites and at larger scales in such complex tropical rainforest would be facilitated from drone-based photogrammetric techniques as well as automatic tree crown detection since ground-based inventory would be labour intensive and difficult for crown assessments. Tree crown detection based on canopy height model (CHM) is common but difficult to implement in dense tropical rainforest. In a recent study, AMS3D (Adaptive MeanShift 3D, a multimodal point-cloud-based ITC detection algorithm), was reported to be suitable for heterogeneous tropical rainforest stands (Ferraz et al. 2016) and to perform better than other ITC detection methods in a lowland tropical rainforest in French Guiana (Aubry-Kientz et al. 2019). Our study aimed at addressing the spatial heterogeneity of canopy transpirations across and among the sites in tropical rainforest by predicting the canopy transpirations using drone-based crown metrics and automatic tree crown detection approach.

On the other hand, oil palm monoculture are more homogeneous and less complex as compared to tropical forest. A biodiversity enrichment experiment, Efforts-BEE, was set up in a commercial oil palm plantation by planting native tree species and establishing oil palm agroforests in-order to alleviate the ecological impacts of oil palm cultivation in Sumatra (Teuscher et al. 2016). Our study also aimed at addressing the E_t heterogeneity between such oil palm agroforest and oil palm monoculture; which may provide a better understanding about the water use patterns of the experimental site.

1.4 Multi-level temporal dynamics

The diurnal pattern of oil palm water use and its influence on environmental drivers is important to understand the water use characteristics. Previous studies observed a pronounce hysteresis while assessing the influences of environmental drivers on oil palm water use; which possibly link to other mechanisms such as stem water storage, stomatal conductivity or hydraulic conductance in oil palm (Niu et al. 2015). Similar studies on oil palm water fluxes also discussed the possible existence of stem water storage in matured oil palms, due to early peaks of sap flux density (Röll et al. 2015). Understanding the temporal dynamics (at higher resolution) of oil palm water use would eventually provide detailed information about such hysteresis and may explain mechanisms such as the role of the stem of water storage in oil palm daily water use. The role of internal water storage have been studied and recognized previously in tropical trees (Goldstein et al. 1998; Meinzer, James, and Goldstein 2004), subtropical trees (Oliva Carrasco et al. 2015), temperate trees (Cermák et al. 2007; Köcher, Horna, and Leuschner 2013); but not yet studied in oil palm (*Elaeis guineensis* Jacq.). While there had been evidence of the importance of internal water storage in the arborescent palm (*Sabal palmetto;* and reported that the stem water storage maintained the leaf water content for 100 days in 4 m tall palm when soil water supply was prevented (Holbrook and Sinclair 1992). Time lag analysis is one of the common approaches where estimates of stem water storage based on the time lag between the sap fluxes at different heights and are a function of the amount of water that is extracted from storage tissues to canopy transpiration losses (Pfautsch, Hölttä, and Mencuccini 2015). Several studies have been compared the time lag duration between the canopy transpiration and base stem sap flow in-order to estimates the stem water storage (Goldstein et al. 1998; Köcher et al. 2013; Phillips et al. 1999). For woody species, time lag depends on buffering capacitances associated between stem water and canopy fluxes (Edwards et al. 1986; Hunt and Nobel 1987) and also positively associated with plant size (Goldstein et al. 1998; Oren, Werk, and Schulze 1986; Phillips et al. 1999). The time lags between stem and branch may also be dependent on anatomical characteristics of the vascular system (Čermák, Kučera, and Penka 1976).

For oil palm, measurements of sap flux techniques were conducted at leaf petioles and the specific parameter was calibrated for oil palm previously (Niu et al, 2015). However, measurements on the stem of the oil palm have not been done and the stem with radial profile consideration yet can explore. In recent times, the radial sap flux gradient in the tree stem has been reported in many studies (Čermák et al. 1992; Delzon et al. 2004; Edwards and Booker 1984; Link et al. 2020; Phillips, Oren, and Zimmermann 1996) and assuming uniform sap flux across the radial direction leads to high errors and uncertainties while estimating whole-tree water use (Čermák and Nadezhdina 1998; Ford et al. 2004; Kumagai et al. 2005). It would be interesting to understand the nature of sap flux across the radial profile of the oil palm stem.

1.5 Outline and Objectives of the study

This study was conducted within the framework of an interdisciplinary project, the CRC 990 (Collaborative Research Centre 990: Ecological and Socioeconomic Functions of Tropical Lowland Rainforest Transformation Systems on Sumatra, Indonesia; www.uni-goettingen/crc990), and Sub-Project A02 ('Tree and palm water use characteristics in rainforest transformation systems'). The study was conducted in Jambi province, Sumatra,

Indonesia, covering sites of oil palm monoculture plantation, oil palm agroforest and tropical rainforest. We conducted sap flux measurements in oil palm monocultures, oil palm agroforest and tropical rainforest using three different types of sap flux methods; according to the study objectives. At the same time, drone multi-copter equipped with the digital camera was used to capture images in all the sites and further used to construct a 3D model of the stand and subsequently, delineate the tree and oil palm crown structure. We extend further to apply automatic crown delineation in forest sites using the AMS3D method. Detail methodology and measurement schemes are provided in the respective chapters.

The main objectives of the study were:

(1) To test scaling approaches from individual plants to stand transpiration using drone-based photogrammetry derived crown metrics

I. To test drone derived crown variables for the prediction of tree and palm water use
II. To analyze uncertainties resulting from scaling plant water use to stand-level transpiration
III. To compare transpiration rates of an oil palm monoculture to an oil palm agroforest.

(2) To analyze spatial variation in plant transpiration

I. To test the relationship between tree water use and crown metrics
II. To predict spatial variability of rainforest canopy transpiration within and across plots, including differences between riparian and upland plots

(3) To assess multi-level temporal dynamics of water circulation in oil palm

I. To assess the radial profile of sap flux density in the stem of oil palm
II. To analyse the hysteresis between sap flux densities and environmental drivers, and
III. To analyse the role of stem water storage

This book comprises five chapters, with the first chapter being a general introduction, chapters two to four comprises three manuscripts (two published and one final draft); which addresses the above-mentioned objectives and final chapter provides synthesize and future scope of this work.

CHAPTER 2

Drone-based photogrammetry-derived crown metrics for predicting tree and oil palm water use

Ahongshangbam, J.[1 *†], Khokthong, W.[1*], Ellsäßer, F [1]. Hendrayanto, H. [2], Hölscher, D. [1, 3], Röll, A. [1]

[1] University of Goettingen, Tropical Silviculture and Forest Ecology, Germany

[2] Bogor Agricultural University, Forest Management, Indonesia

[3] University of Goettingen, Centre of Biodiversity and Sustainable Land Use, Germany

* The authors contributed equally to the manuscript

† Correspondence to: Joyson Ahongshangbam, Tropical Silviculture and Forest Ecology, Georg-August-Universität Göttingen, Büsgenweg 1, 37077 Göttingen, Germany. E-mail: jahongs@gwdg.de Telephone: +49 (0) 551 39 12101; Fax: +49 (0)551 394019

Published in ***Ecohydrology*** (2019) 12:e2115, doi: https://doi.org/10.1002/eco.2115

2.1 Introduction

Transpiration is a central flux in the ecosystem water cycle. In forests or similar vegetation types, it is often estimated from individual plant water use assessments, for example with sap flux techniques (Granier, 1985; Wullschleger, Meinzer, & Vertessy, 1998). In most studies, the number of plants directly analyzed for water use is lower than the number of plants in the stand. The individual plant water use rates are then scaled to stand-level transpiration by biometric variables. Scaling is thus a critical issue that needs to be optimized in order to improve transpiration estimates and to reduce associated uncertainties (Hatton & Wu, 1995; Jarvis, 1995; Moore, Orozco, Aparecido, & Miller, 2017).

Candidate variables for scaling include tree diameter, crown metrics and leaf area. Among these, tree diameter and the number of trees (stand density) are often used, as they are easy to assess and often available from forest inventories. The relationships between tree water use and tree diameter often have R^2 values around 0.66 (Yue et al., 2008; Schiller, Cohen, Ungar, Moshe, & Herr, 2007), but closer (Wang, Xing, Ma, & Sun, 2006) and less close correlations (Kume et al., 2009) have also been observed. Stem diameter has some limitations that include a potentially slow response to concurrent dynamics in the stand such as crown damages or crown expansions into gaps. In addition, recently increasingly monocot species such as bamboos and palms came into the focus of transpiration studies (Röll et al., 2015; Mei et al., 2016), in which intra-specific diameter variation may be low but nonetheless variation in water use occurs. Leaf area index can be a very powerful variable for scaling (Hatton & Wu, 1995; Vertessy, Benyon, O'Sullivan, & Gribben, 1995; Medhurst, Battaglia, & Beadle, 2002), but it is often only available at the stand level and not at the tree level. In contrast, crown dimensions are easier to measure and thus more commonly available and yielded good results in mature oak (*Quercus robur*) forest (Čermák, 1989). Similarly, in *Taxodium distichum* forest and olive orchard, crown structure correlated closely with tree water use (Oren, Phillips, Ewers, Pataki, & Megonigal, 1999; López-Bernal, Alcántara, Testi, & Villalobos, 2010). Crown exposure also indirectly affected transpiration by influencing leaf wetness and dryness in a premontane forest of Costa Rica (Aparecido, Miller, Cahill, & Moore, 2016).

Despite the long-recognized potential of crown variables for scaling up from tree water use to stand transpiration, diameter based approaches remain popular, as crown variables are more difficult and time consuming to assess in ground-based stand inventories. With the recent development of drone technologies and their application in ecological studies this might change. Drones equipped with optical detectors such as cameras capturing specific light wave lengths or laser-based approaches offer new opportunities for crown and canopy assessments (Díaz-Varela, de la Rosa, León, & Zarco-Tejada, 2015; Thiel & Schmullius, 2016; Barnes et al., 2017). Crown variables such as crown length (Kallimani, 2016), crown diameter (Lim et al., 2015; Panagiotidis, Abdollahnejad, Surový, & Chiteculo, 2016) or crown volume (Torres-Sánchez, López-Granados, Serrano, Arquero, & Peña, 2015) were calculated using photogrammetric techniques. Even though drone technologies have previously been applied in ecohydrological studies (Vivoni et al., 2014), the applicability of drone-based photogrammetry for scaling up tree water use to stand-level transpiration has to our knowledge not yet been explored.

Uncertainties associated with sap flux measurements and stand level estimates of transpiration are manifold and include the assessment of sap flux variation in a given tree, the number of trees sampled, and the scaling (Peters et al., 2018). For a better understanding of ecohydrological consequences with land-use and land-cover change, it will be important to produce stand-level transpiration estimates with a high accuracy and thus, a low associated uncertainty. The basis for this is the further optimization of current sampling and scaling schemes, potentially also by employing innovative drone-based methods.

In our study, we assessed relationships between crown metrics and the water use of oil palms and trees in lowland Sumatra, Indonesia. In this region, natural forests have largely been converted and monoculture oil palm plantations are widespread (Drescher et al., 2016). The land cover change and the expansion of oil palm plantations are associated with losses of biodiversity and impaired ecosystem functions (Barnes et al., 2014; Clough et al., 2016; Dislich et al., 2017). Transpiration rates from commercial oil palm plantations can be high and may exceed those of remaining forests (Röll et al. 2015, Meijide et al. 2018). To test possibilities of alleviating the ecological impacts of oil palm cultivation, a biodiversity enrichment experiment, Efforts-BEE, was set up in a commercial oil palm plantation by planting native tree species and establishing oil palm agroforests (Teuscher et al., 2016). Within Efforts-BEE, we conducted our study on plant water use and scaling by crown variables. The objectives were (1) to test drone derived crown variables for the prediction of tree and palm water use, (2) to analyze uncertainties resulting from scaling plant water use to stand-level transpiration, and (3) to compare transpiration rates of an oil palm monoculture to an oil palm agroforest.

2.2 Methods

2.2.1 Study area and sites

The study was conducted in Jambi province, Sumatra, Indonesia. The region is tropical humid, with mean annual precipitation of 2235 mm yr^{-1} and average annual temperature of 26.7° C (Drescher et al., 2016). The study sites were located just south of the equator (01.95° S and 103.25° E), within the commercial oil palm plantation PT Humusindo, near Bungku village. Mean elevation is 47 m asl. The biodiversity enrichment experiment (EFForts-BEE) was established in monoculture oil palm plantations. Oil palms were planted in a 9 m x 9 m triangular grid resulting in approx. 143 oil palms per hectare; the age of the oil palms at the time of study was approx. 9-15 years (Teuscher et al., 2016). The broad age range refers to the entire experiment with 56 plots that covers an area of about 150 ha. After thinning of oil palms, six native tree species were planted in a 2 m x 2 m grid. The tree species were mixed in a way to maximize the number of hetero-specific neighbors (i.e. no con-specific rows or groups) (Teuscher et al., 2016). There are 52 experimental plots varying in plot size and in tree species diversity level. In addition, there are also 4 control plots with oil palm management as usual, and no enrichment planting. Our main study site was at a 40 m by 40 m plot with six tree species planted (figure 1) and a nearby monoculture control plot of the same size. The agroforest plot was selected based on the criteria plot size (as big as possible, i.e. 40 m by 40 m) and highest tree diversity level (six tree species). The monoculture control plot was located approx. 60 m away from the agroforest plot. At the selected agroforest and monoculture study plot, oil palms were of similar age. In the agroforest, the studied oil palms had an average meristem height of 6.8 ± 0.2 m (mean ± SD), while the sample trees had an average height of 4.7 ± 0.6 m (Appendix A). The reported measurements were conducted between September and November 2016, which was the beginning of the rainy season.

Figure 1: Aerial view of a studied oil palm agroforestry plot. Three years prior to the study, the stand was thinned with reduction in oil palm stems by 40% and six tree species were planted

2.2.2 Sap flux measurements

Eight palms and 16 trees were equipped with sap flux sensors. Selected tree species were *Archidendron pauciflorum*, *Parkia speciosa*, *Peronema canescens* and *Shorea leprosula*. As *Shorea leprosula* did not perform well on the multi-species plot, it was measured on a nearby single tree species enrichment plot, under otherwise very similar conditions. One further tree species, *Dyera polyphylla*, was not included in the measurements because almost all individuals had died on the multi-species plot and no plot with well performing *Dyera polyphylla* trees was available nearby. *Archidendron pauciflorum, Parkia speciosa* and *Peronema canescens* are early successional and light demanding species (Aumeeruddy, 1994; Lee, Wickneswari, Clyde, & Zakri, 2002; Orwa et al., 2009; Lawrence, 2001); *Shorea leprosula* is considered a gap opportunist (Ådjers, Hadengganan, Kuusipalo, Nuryanto, & Vesa, 1995; Bebber, Brown, Speight, Moura-Costa, & Wai, 2002).

Sap flux sensors were installed in four trees for each tree species and on four oil palms in an oil palm agroforest, and additionally on four oil palms in the oil palm monoculture.

For trees, we used heat ratio method sensors (HRM, Burgess et al., 2001; ICT International, Australia). One HRM sensor per tree was installed radially into the xylem at breast height. To process raw data we used the software Sap Flow Tool, version 1.4.1 (ICT International, Australia). The mean sap velocity output data was converted into 'sap flow' (cm^3 h^{-1}) by multiplying it with the cross-sectional water conductive area A_c (cm^2). As the studied trees were rather small (diameter at breast height, DBH < 11 cm), we considered A_c to be equal to the cross-sectional area at breast height. Estimation errors associated with assuming fully conductive cross-sectional areas of the relatively small trees for the up-scaling to tree water use are likely to be small; for similar sized trees Delzon, Sartore, Granier, & Loustau (2004) found a difference of approx. 4% with this assumption.

For oil palms, we used thermal dissipation probes (TDP, Granier, 1985) as this method had previously been tested on oil palm and a sampling scheme had been developed (Niu et al., 2015), which we followed closely. Like Niu et al. (2015), we installed the TDP sensors in leaf petioles rather than the stem of oil palms due to presumably higher vessel density and homogeneity in vascular bundle distribution (Madurapperuma, Bleby, & Burgess, 2009; Renninger, Phillips, & Hodel, 2009). Niu et al. (2015) also tested the influence of leaf characteristics such as leaf orientation, inclination and horizontal shading on leaf water use for 56 oil palm leaves, but no statistically significant effects were observed. The authors argued that the examined factors partly counteract (Niu et al. 2015). We followed their suggested scheme in our study and selected four leaves per palm in the cardinal directions. Sap flux density J_s (g cm^{-2} h^{-1}) was calculated using the equation derived by Granier (1985), but with oil palm specific, calibrated equation parameters (Niu et al., 2015). Zero-flux conditions were examined following Oishi et al. (2008); it was found that zero-flux conditions were met during the early morning hours during our entire sap flux measurement period. Individual leaf water use rates (kg day^{-1}) were calculated by multiplying J_s daysums by A_c of the according leaf petioles. Those were derived from a previously presented linear relationship between petiole baseline length (which was measured with a caliper) and A_c at the location of the sensor (Niu et al. 2015). Individual daily leaf water use rates were averaged for each palm and multiplied by the number of leaves per palm to derive palm water use rates (kg day^{-1}). Water use rates were based on averages of three sunny days on which soil moisture was non-limiting in order to minimize the effects of varying environmental conditions; this approach is in accordance with previous research on oil palm water use (e.g. Hardanto et al., 2017; Niu et al., 2015; Röll et al., 2015). In the nomenclature across the applied sap flux methods, we follow Edwards, Becker, & Cermák (1997) in expressing individual tree and oil palm water use as mass per time (kg day^{-1}) and stand-scale transpiration in 'mm day^{-1}'.

2.2.3 Drone image acquisition and processing

At the time of the sap flux measurements, drone flights were conducted using an octocopter (MikroKopter OktoXL, HiSystems GmbH, Germany) equipped with a digital RGB camera (Nikon D5100, Japan). Flight routes were planned with MikroKopter-Tool V2.14b. Flight altitude was 39 m above ground, flight speed was 7.2 km h^{-1} and one picture was taken per second (Appendix B).

The flight missions were performed in circular and grid pathways to get different perspectives and an overlap of 70% for the construction of 3D maps. After eliminating blurry pictures, 3D point clouds were created from an average of 600 geo-referenced images per study site with Agisoft Photoscan Professional 1.2.6 software (Agisoft LLC, Russia). The achieved point cloud density was 3 points cm^{-2}. In the analysis, we used the pictures from one single flight to construct the 3D models.

The workflow included image alignment, georeferencing, building dense point clouds, the generation of digital elevation models (DEM) and orthomosaic generation. Ground-control points printed as 8-Bit barcodes and laid out during the flight campaigns were used to determine the overall positional accuracy of orthomosaic images. The 3D point clouds were generated using the Structure from Motion (SfM) technique (Westoby, Brasington, Glasser, Hambrey, & Reynolds, 2012; Lowe, 2004). Orthomosaic and digital elevation models (DEM) were created for each plot for further visualization and interpretation.

In order to create canopy height models (CHM), digital terrain models (DTM) were generated from the point cloud data. For this, the three main parameters (maximum angle, maximum distance and cell size) were defined with Agisoft's ground point classifier tool and used to differentiate ground and non-ground points. The classified ground points were converted to raster format as DTM. Further, we overlaid the DEM and DTM and applied smooth filters to derive the canopy height model. Subsequently, crown polygons were delineated for target trees and oil palms through visual interpretation and tree location information. One major challenge of using aerial imagery for delineating individual tree canopies is the overlapping of crowns. It was not major issues in our study as the studied trees are young and located in gaps created by the previous thinning of oil palms. The 3D crown models of the studied palms and trees (extracted from the SfM point clouds) were derived from multiple shots at different angles and positions, thus allowing to delineate even overlapping canopies. Additionally, the very high point cloud density of 3 points cm^{-2} allowed modeling the crown structures in great detail. However, for some sample trees we experienced difficulties with automatic 3D segmentation, e.g. when branches from different trees connect (Tao et al. 2015). In such a case, we performed additional manual segmentation and processing and added clusters for the automatic approach (Trochta, Kruček, Vrška, & Kraâl, 2017). The individual canopy height of trees and meristem height of oil palms were obtained by overlaying individual crown polygons with the CHM. For trees, the highest point in CHM within the individual crown polygon was considered as the canopy height of trees (Birdal, Avdan, & Türk, 2017), while the lowest point was taken as the meristem height of oil palms. As a ground-based reference, canopy height of each individual was measured using a pole, and canopy width and projection area were established with the vertical sighting method (Preuhsler 1979, also see Pretzsch et al. 2015) in the eight cardinal directions. The heights obtained by the drone-based and the ground-based methods were well correlated along a 1:1 line ($R^2 = 0.69$, $P < 0.001$; Appendix C). Also, the canopy diameter obtained by terrestrial measurements and drone based analyses were highly correlated along a 1:1 line ($R^2 = 0.95$, $P < 0.001$), suggesting the applicability of the drone based approach. The PolyClip function in Fusion software v3.6 (USDA, USA) was used to extract individual point clouds for each tree and oil palm crown. Crown variables of each individual were obtained using measurement marker functions in the same software. For crown volume and planar area, the point clouds were interpolated in R software v3.4.3 (R Development Core team, 2016) using the Alphashape3D (Lafarge & Pateiro-Lopez, 2014) and rLiDAR (chullLiDAR2D, Silva et al., 2017) packages, respectively.

There are several different ways to compute crown volumes including convex hull and alpha shape algorithms (Colaço et al., 2017). In convex hull, it constructs an envelope by considering the number of input points belongs to the convex hull to represent the outward curving shape of tree crowns. In the alpha shape approach, a predefined and reduced alpha value serves as size criterion to construct more details, thus shrinking the corresponding convex hull closer down to the 3D point cloud (Pateiro-Lopez and Rodriguez-Casal, 2010; Colaço et al., 2017). In our study, we calculated the crown volumes for both trees and oil palms with a convex hull algorithm and alpha shape algorithms, the latter using the alpha values 0.75, 0.50 and 0.25 (Appendix D). Two contrasting models (convex hull and alpha shape 0.25) are illustrated in figure 2 for a studied oil palm and a studied tree.

2.2.4 Statistical analyses

To test for differences in tree water use among species, and for differences in oil palm water use between oil palm agroforest and oil palm monoculture, we used ANOVAs, followed by Posthoc Tukey's HSD; differences were assumed as significant at $P < 0.05$.

Plant size related variables such as crown volumes as predictor of plant water use were tested by linear regressions. We tested the variance of residuals for normal distribution by the Shapiro-Wilk test and homoscedasticity with residual plot analysis. The null hypothesis of normality was rejected at $P < 0.05$.

The linear regressions served as the basis for subsequent scaling of tree- and palm-level water use to stand-level transpiration. To compare the uncertainties associated with different scaling variables, we performed parametric bootstrapping with the linear relationships between water use and the predictor variables with 50,000 iterations using the R package 'boot' (Canty & Ripley., 2017; Davison & Hinkley, 1997). This yielded estimates of means and corresponding standard deviations as measures of uncertainty.

All statistical analyses and plotting were performed with R version 3.4.3 (R Development Core team, 2016).

Oil palm | **Tree: *Shorea leprosula***

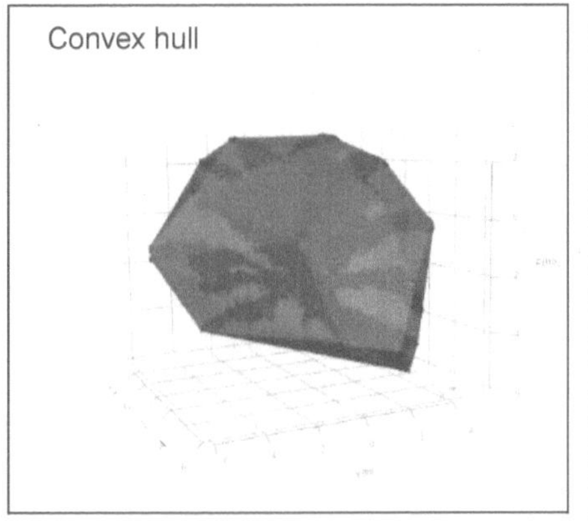

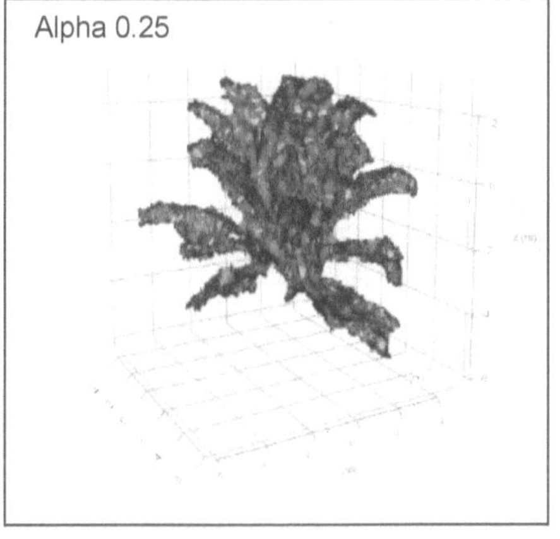

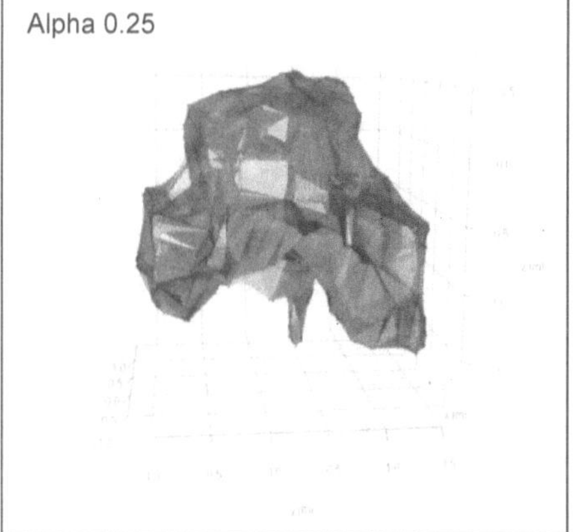

Figure 2: Canopy of an oil palm and a tree (*Shorea leprosula*) using point clouds from the drone missions and convex hull and alpha shape algorithms. Other tree species are shown in appendix D.

2.3 Results

2.3.1 Plant water use

On sunny days, the daily water use per palm ranged between 158 and 249 kg day^{-1}, and on average was by 32% higher in the agroforest than in the monoculture (ANOVA, $P < 0.01$). Daily water use of the inter-planted trees was much lower and per tree ranged from 1.1 to

19.8 kg day^{-1}. There were species-specific differences among the trees ($P < 0.001$) (Appendix A).

2.3.2 Drone-derived crown metrics and their relation with plant water use

Crown volumes (convex hull) for the eight oil palms with sap flux measurements ranged between 332 and 831 m^3, and on average were by 79% higher in the agroforest than in the monoculture. Crown volumes (convex hull) of the trees were much lower and ranged between 0.95 and 81.0 m^3. There were species-specific differences among the trees (Appendix A).

Crown metrics were highly correlated with tree and palm water use (table 1). For oil palm, crown volume convex hull explained 69% of the observed palm-to-palm variability in daily water use ($P = 0.01$). For trees, crown volume models with an alpha level 0.25 (see Appendix D) explained 81% of tree-to-tree variability ($P < 0.001$) (figure 3) across the studied species. Due to violated quality criteria (Shapiro-Wilk test, $P = 0.000042$), there was however no single linear crown volume model that fit both oil palms and trees. Nonetheless, the single linear relationship crown volume alpha 0.75 to tree/palm water use is depicted in Appendix E.

For trees, stem diameter as measured in ground-based inventories explained 65% of the variability observed in daily tree water use ($P < 0.01$), while for oil palms no significant ground-based explanatory variables were available for comparison

2.3.3 Transpiration estimates and uncertainties

Based on scaling with crown volumes, the stand-level transpiration estimate in the oil palm agroforest is 1.9 mm day^{-1} and 3.0 mm day^{-1} in the oil palm monoculture (table 2). Scaling with ground-based DBH measurements in trees resulted in only minor differences in stand transpiration estimates. For trees, bootstrapping suggests that the estimate based on crown volume is associated with an uncertainty due to scaling of 28%. In contrast, using diameter for scaling results in an uncertainty of 100%. For the oil palms in the agroforest and the monoculture, the uncertainty estimates associated with crown volume scaling were 37% and 35%, respectively.

Table 1. Linear regressions between daily water use (kg day^{-1}) and different aerial and ground based variables of oil palms (n=8) and trees (n=15). Only those linear regressions which satisfy normality and homoscedasticity conditions are presented.

			Equation b_0 **- water use** b_1 **- variables**	**P value**	R^2
Drone based					
Crown volume (m^3)	Oil palms	convex hull	$b_0 = 0.14\ b_1 + 122$	$P = 0.010$	0.69
		alpha 0.75	$b_0 = 0.74\ b_1 + 49.1$	$P = 0.038$	0.53
	Trees	alpha 0.75	$b_0 = 0.39\ b_1 + 2.12$	$P < 0.001$	0.73
		alpha 0.5	$b_0 = 0.51\ b_1 + 1.84$	$P < 0.001$	0.77
		alpha 0.25	$b_0 = 0.82\ b_1 + 1.70$	$P < 0.001$	0.81
Ground based					
DBH (cm)	Oil palms		-	-	-
	Trees		$b_0 = 2.46\ b_1 - 8.42$	$P < 0.01$	0.65

Table 2. Transpiration of the oil palm agroforest and monoculture plot with uncertainties for the scaling from individual plants to the plot level by bootstrapping linear relationships. For uncertainty estimates from ground-based scaling in oil palm we used an approach by Niu et al. (2015), which is based on the number of leaves that measurements were performed on and the resulting cumulative coefficient of variation (marked with an *).

		Transpiration (mm day^{-1}) Estimate ± uncertainty
Drone-based		
Agroforest	Trees	0.28 ± 0.08
	Oil palms	1.61 ± 0.61
	Total	1.89 ± 0.69
Monoculture	Oil palms	3.04 ± 1.05
Ground-based		
Agroforest	Trees	0.38 ± 0.38
	Oil palms	1.67 ± 0.88*
	Total	2.05 ± 1.26
Monoculture	Oil palms	2.96 ± 0.78*

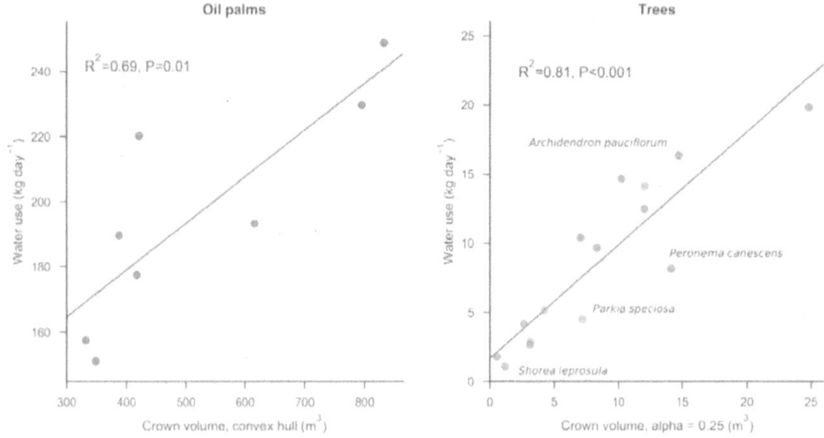

Figure 3: Daily water use of (a) oil palms and (b) trees versus crown volumes. Note the different crown volume models and scales

2.4 Discussion

In our study, we found that drone-based assessments of oil palm and tree crowns predicted individual plant water use quite well and better than e.g. diameter in trees, and thus led to reduced uncertainties in spatial scaling and stand-level estimates of transpiration.

A popular variable for the prediction of plant water use is stem diameter. In our study, DBH yielded an R^2 of 0.65, which is quite similar to several recent studies (Yue et al., 2008; Schiller, Cohen, Ungar, Moshe, & Herr, 2007; Granier, Biron, & Lemoine, 2000). In our study as in many others, the relationship between DBH and tree water use was found to hold across species. In contrast, in a premontane forest in Costa Rica the correlation of water use to DBH showed differences among species (Moore et al., 2017). Likewise, species-specific trajectories were suggested from reforestations in the Philippines (Dierick & Hölscher, 2009). There are further general concerns in using diameter for scaling. As such, diameter integrates over large time spans and a tree may have achieved its diameter under conditions that no longer prevail at the time of study. Cases in point are damages by storm or lightning, or in the other direction crown expansion into a gap that was formed by the dieback of a neighbor.

Our study also included oil palm, a monocot plant which lacks secondary diameter growth. Consequently, significant correlations between stem diameter and water use can hardly be expected. Thus far, to our knowledge no scaling scheme from an individual oil palm to the stand level had been established. Based on leaf level measurements in 56 oil palm leaves, Niu et al (2015) tested for relationships between leaf characteristics (e.g. orientation, inclination, horizontal shading) and leaf water use but did not find significant relationships. In contrast, the approach of our study with crown volume and whole plant water use resulted in an R^2 of 0.69 ($P = 0.01$). Based on their results, Niu et al. (2015) suggested a non-stratified sampling scheme. Our results would suggest that a sampling scheme in oil palm would benefit from representing different crown dimensions.

For trees and palms the best fitting (as based on high R^2 and low P) crown volume model differed with alpha 0.25 for trees and convex hull for palms. There was however one single intermediate crown volume model, alpha 0.75, that appears suitable for both trees and palms (table 1). However, applying this model for the pooled dataset of all trees and palms resulted in non-normality and too high heteroscedascity to be accepted (Shapiro-Wilk test, $P = 0.000042$), even though R^2 was very high and the P value was low (Appendix E). Our dataset certainly lacks values in the mid-range of crown volume and water use for a further examination of this 'universal' crown model. Also, it can be seen that crown alpha 0.75 is not the best predictor for oil palm water use. However, the universal model may indicate that trees and oil palms do not differ significantly in water use per crown volume, even though more and more equally distributed data will be needed to further test this contention. On the other hand, it may also well be that a universal crown volume to plant water use relationship does not exist. As such, across (tree) species, e.g. when comparing early successional and late successional species, substantial differences regarding crown shape, the occurrence of sun and shade leaves and leaf stomatal conductance exist (Bazzaz, 1979; Poorter, Bongers, & Bongers, 2006).

Sap flux measurements and subsequent scaling up to the stand level are associated with a multitude of uncertainties, including the positioning and number of a sensor in a given plant, methods of zero-flow conditions and sensor calibration (Peters et al., 2018), as well as the number of plants studied. Our study addresses the spatial scaling from the individual plant to the stand. The uncertainty estimates as the result of the applied bootstrapping are directly related to the explained variance in the linear relationships with water use. They suggest that for trees the uncertainty of the stand-level estimate is 28% with drone-based imagery, whereas it is 100% with ground-based diameter measurements. The drone-based approach thus has at least one clear advantage. For oil palms, our reported uncertainty of 37% is the first estimate that to our knowledge addresses whole-plant to stand scaling. However, Niu et al. (2015) estimated that counting leaves per oil palm and oil palms per stand, and scaling based on sap flux measurements in 12 leaves, would result in uncertainty of stand-level transpiration of 14%. For oil palms it thus seems that the previously proposed ground-based method has an advantage. Nonetheless, the crown dimension approach is still valuable, as it may also allow to estimate water use across different conditions. For example, in our case an oil palm stand was thinned and trees were inter-planted in gaps three years prior to the study (agroforest), whereas the control stand remained untreated (monoculture). We found significant differences in crown volume and water use of the studied oil palms, but the two variables were significantly related to each other across treatments. The ground-based leaf-count approach, on the other hand, was previously only tested in one single stand with homogenous conditions. Their applicability will have to be tested further in follow-up studies focusing on how to best assess (and reduce) such estimation uncertainties.

The crown volumes in our study were derived from RGB images and a photogrammetric approach. Other drone-derived structural variables such as height and projected crown area show a high correlation with ground-based reference measurements along a 1:1 line, suggesting the applicability of the aerial method. The point cloud density in our study was 3 points cm^{-2}, which can be regarded as quite high and compares to or is even higher than those that result from laser scanning (Vauhkonen, Næsset, & Gobakken, 2014). Drone-based imagery performs particularly well for the upper part of the canopy, which is also where a large part of the transpiration takes place. So far, we only tested this method in a relatively simply structured monoculture and an oil palm agroforest with relatively young trees. As we regard the results as promising, it will be interesting to test it in more heterogeneous stands in next step.

Oil palm water use in the studied monoculture and the agroforest ranged between 158 and 249 kg day^{-1}. The studied monoculture is relatively intensively managed, with fertilizer application including 230 kg N ha^{-1} $year^{-1}$ (Teuscher et al., 2016). The observed water use rates exceed those of small-holder plantations of similar age (108 ± 8 kg day^{-1}, mean ± SE among eight sites) and compare to values from another intensively managed, commercial oil palm monoculture plantation in the region (178 ± 5 kg day^{-1}) (Röll et al., 2015; Meijide et al., 2018). Thus, our data indicates that intensive oil palm management leads to high water use rates.

The water use per oil palm in the agroforest was 31% higher than in the monoculture. This is likely due to the reduction of oil palm stand density by previous thinning in the agroforest, which leads to increases in light, soil water and nutrient availability for the remaining oil palms in the stand. This is also in line with a previous study showing 36% higher per-palm fruit yield in thinned agroforests than in untreated monocultures (Gérard et al., 2017). The mean individual tree water use in agroforest, on the other hand, was very low (1.1 - 19.8 kg day^{-1}) compared to the water use of the surrounding oil palms. The large difference in tree water use is likely due to the substantial differences in tree size (4.2 cm vs 11 cm) and canopy volume (1.1 m^3 vs 24 m^3). However, tree size also coincides with species identity in our case, so 'ultimate reasons' cannot be disentangled. However, these low absolute rates of the inter-planted trees of relatively small-diameter (DBH range 4.2 – 11.0 cm) compare well to values provided for rubber trees of similarly small diameter in a previous study in the lowlands of Sumatra (Niu, Röll, Meijide, Hendrayanto, & Hölscher, 2017). The general observation of high water use per palm also corresponds with data from Amazonian fruit plantations, where it was found that palms consumed 3.5 times more water than trees (Kunert, Aparecido, Barros, & Higuchi, 2015).

Scaled to the stand-level based on our aerial approach, stand transpiration of the oil palm agroforest (1.9 mm day^{-1}) was 37% lower than in the oil palm monoculture (3.0 mm day^{-1}). The higher per-palm water use in the oil palm agroforest thus did not compensate for the reduction in oil palm stand density when scaled to the stand level. The 3-year old, comparably small inter-planted trees in the agroforestry plot contributed rather little to overall stand transpiration (15%). The oil palm agroforestry experiment EFForTS-BEE was designed and established to test possibilities of reducing the impact of oil palm cultivation on biodiversity and ecosystem functioning. Oil palm monocultures are associated with ecohydrological problems arising from high transpiration rates and low soil water infiltration capacities (Merten et al., 2016). At the time of study, transpiration rates from the agroforest were substantially reduced in comparison to the commercial monoculture, which may help to alleviate some of the ecohydrological problems. However, restoring the integrity of the local hydrological cycle by means of oil palm agroforestry will also largely depend on whether soil infiltration capacities will increase due to the presence of the inter-planted trees.

2.5 Conclusions

Crown volumes derived from drone-based imagery predicted tree and palm water use quite well. For oil palms, such a scaling variable at the whole-plant level was previously not available. For predicting individual water use, tree crown volumes performed better than the more conventionally used variable stem diameter. In consequence, stand-level transpiration estimates based on crown volumes were associated with reduced uncertainties. We therefore see great potential for future applications of our aerial method in studies scaling plant water use from individual plants to the stand level.

3.1 Introduction

Tropical rainforests comprise a complex 3D structure, rich tree species diversity and encompass heterogeneous site conditions [1,2]. Transpiration (E_t) is a central flux in hydrological cycles and contributes to cloud formation, turbulence and atmospheric cooling [3], and is thus an ecosystem service related to climate regulation. The prediction of canopy E_t by tropical rainforests including its spatial heterogeneity may be fostered by a better understanding of the linkage between structure and function and forest structure variability across sites.

Rainforest E_t can be derived from sap flux measurements in individual trees. Therein, tree-level E_t is scaled-up to the stand-level by using allometric relationships with stand inventory variables such as tree diameter at breast height (DBH) [4]. Due to often relatively high unexplained variability in the DBH to water use relationship, resulting uncertainties of E_t at the stand-level are also relatively high. In a tropical agroforest, Ahongshangbam et al. [5] found that drone-derived crown metrics correlated much better with tree water use than DBH. In consequence, uncertainties associated with the scaling to stand-level E_t were reduced considerably. However, the studied stands were relatively simply structured and the trees were small in stature. The reported crown metrics vs. water use relationship cannot be applied to other vegetation types such as more heterogeneous tropical forest without further testing. Airborne tree crown assessments are also potentially promising for reducing E_t scaling uncertainties in tropical forests, but there are no studies confirming this yet.

The spatial heterogeneity in rainforest E_t is potentially related to variability in site conditions. In North-America, several upland-to-wetland gradients were analysed in order to evaluate the significance of site conditions for tree and stand E_t [6–8]; pronounced differences in E_t were observed and it was concluded that it is necessary to include plots in different topographic positions for landscape-level assessments. For tropical rainforest regions, such studies are rare, but the influence of the water table on E_t of certain species was analysed in northern Australia [9] and on Hawaii [10]. In lowland Sumatra, topography and flooding resulted in differences in E_t between upland and riparian oil palm and rubber tree stands [11].

Rainforest structure assessments with conventional ground-based techniques face difficulties in reliably estimating key variables such as crown dimensions, which influence forest-atmosphere water exchange. Drones equipped with LiDAR [12,13] or optical cameras [14,15] appear more suitable for crown assessments. The latter produce high resolution images, from which 3D point clouds can be computed with the Structure from Motion technique (SfM) [16–18]. Once a certain crown assessment methodology is established, relatively large areas can be assessed in a short time [19–21].

Rainforest canopy heterogeneity analyses for predicting E_t across sites and at larger scales would benefit from an automated delineation of individual tree crowns. Individual tree crown (ITC) detection algorithms often use canopy height model (CHM) based tree segmentations derived from local maxima in the CHM [22–24]. However, CHM-based approaches have limitations in dense stands and multi-layered forests as they tend to merge crowns and fail to detect understory trees with narrow crowns [25,26]. More recently, ITC detection based on 3D point clouds showed promising results, with more accurate tree segmentation in intermediate canopy strata compared to CHM-based approaches [27,25,28]. However, many of these studies were carried out in boreal and temperate forests [29,30] which tend to be less complex in structure than tropical rainforests. In a recent study AMS3D (Adaptive MeanShift 3D), a multimodal point-cloud-based ITC detection algorithm, was reported to be suitable for heterogeneous tropical rainforest stands [25] and to perform better than other ITC detection methods in a lowland tropical rainforest in French Guiana [31]. AMS3D was further reported to be able to detect even suppressed or smaller trees with narrow canopies [32].

The present study was conducted in the Harapan rainforest in the lowlands of Sumatra, Indonesia. The forest landscape is undulating with upland and riparian regions. We conducted sap flux measurements and drone-based crown and canopy assessments at four upland and four riparian forest plots. The objectives of our study were (1) to test the relationship between tree water use and crown metrics, and (2) to predict spatial variability of rainforest canopy E_t within and across plots, including differences between riparian and upland plots.

3.2 Materials and Methods

3.2.1 Study region and sites

The study was conducted in the lowlands of Jambi province, Sumatra, Indonesia (Figure 1). The region is tropical humid, with a mean annual precipitation of 2235 mm yr^{-1} and an average annual temperature of 26.7° C [33]. The study sites were located in the

Harapan rainforest, approx. 50 km south-west of the province capital Jambi. The Harapan rainforest was previously selectively logged but is now a protected area [33]. The region is characterised by mixed dipterocarp-dominated lowland rainforest [34]. A previous assessment in the Harapan rainforest covering four upland study plots (2500 m^2 each) found a total of 201 tree species with a DBH $\geq$ 10 cm [35]. The terrain is undulating, dividing the landscape into upland and riparian valley sites. The soil characteristics at upland and riparian sites are sandy loam Acrisols [36] and acidic clay-loam Stagnosols [37], respectively.

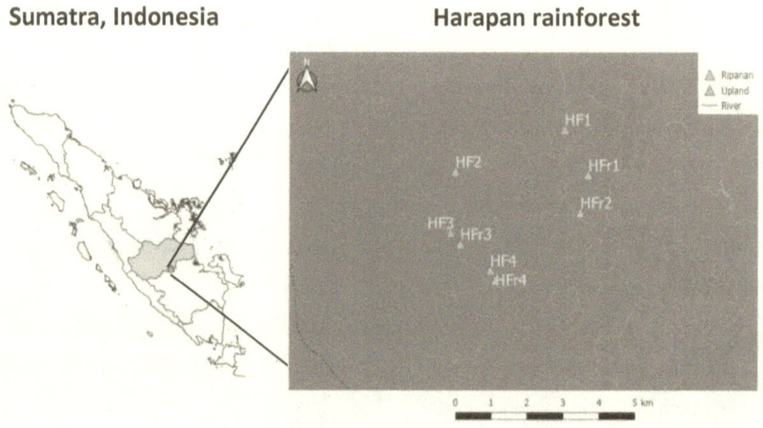

Figure 1. Location of the four upland and four riparian study plots in Jambi Province in the lowlands of Sumatra, Indonesia.

3.2.2 Study plots and stand characteristics

Four study plots were established at upland sites (plot codes HF1, HF2, HF3, HF4) and four at riparian sites (HFr1, HFr2, HFr3, HFr4) within the Harapan rainforest as part of the EFForTs project [33]. The plots were 50 x 50 m^2 in size. Mean elevation at the upland and riparian plots is 65 m and 52 m asl., respectively. At upland plots, the mean tree height was 21.8 ± 0.8 m; at riparian plots, the mean tree height was 18.9 ± 0.8 m (for trees with a DBH $\geq$ 10 cm; Table 1). Among the upland plots the mean nearest distance to neighboring plots was 2.1 km, among riparian plots it was 1.7 km. The mean nearest distance between upland and riparian plots ('plot pairs') was 0.3 km, only for one pair the distance was larger (2.5 km) (Figure 1).

Table 1. Trees per plot, diameters at breast height (DBH, ≥ 10 cm) and tree heights at the eight study plots (50 x 50 m^2).

Plot	Trees per plot (n)	DBH (cm)			Height (m)		
		mean	min	max	mean	min	max
HF1	125	21.9	10.1	67.9	19.7	8.3	52.2
HF2	172	20.1	10.4	67.5	18.4	7.8	48.0
HF3	146	22.6	10.2	80.2	21.1	4.3	44.5
HF4	143	22.5	10.0	76.8	21.0	2.5	48.2
HFr1	135	17.8	10.0	67.0	16.9	1.1	42.5
HFr2	136	20.1	10.0	56.3	18.3	1.4	34.4
HFr3	160	17.9	10.1	51.0	19.9	5.2	36.7
HFr4	140	20.7	10.1	108.1	20.4	4.6	44.0

3.2.3 Sap flux measurements

To assess tree water use, we measured sap flux with thermal dissipation probes (TDP) [38]. We selected one upland plot (HF2) and all four riparian plots (HFr1, HFr2, HFr3, HFr4) for these measurements; the remaining three upland plots had already been measured in a previous sap flux study [39]. The three plots not studied with sap flux methods served as independent tests for stand-scale transpiration derived from two different methods. In each plot, 15 trees were equipped with two TDP sensors each, with the exception of HFr1, where only 12 trees were equipped due to a lack of sap flux instrumentation, as this plot was measured at the end of the field campaign. Installation of sap flux sensors and calculation of sap flux density (J_S, g cm^{-2} h^{-1}) followed the methods described in [39] for lowland rainforest in the same region; therein, the original Granier's [38] equation for deriving J_S was applied. Nighttime zero-flux conditions, as described in Oishi et al. [40], were met during the early morning hours. Tree-level water use was derived by considering the water conductive area (A_C, cm^2) of trees and radial sap flux patterns measured by the heat field deformation technique [39].

3.2.4 Remote sensing

3.2.4.1 Drone image acquisition

Drone flights were conducted within the sap flux measurement period between August and December 2016. An octocopter drone (MikroKopter EASY Okto V3, HiSystems GmbH, Germany) equipped with a RGB camera (Sony Alpha 5000 with Sony E PZ 16-50mm lens) was used to capture the images. The drone was additionally equipped with GPS (MKBNSS V3 GPS/Glonass, HiSystems, Germany); the accuracy of the GPS measurements was ± 5 m. Flight routes were planned with MikroKopter-Tool V2.14b and the flight path followed superposing circular and grid patterns. Images were taken at an altitude of 80 m above ground (i.e. 30 - 40 m above canopy). Further flight specifications are provided in Table A1.

3.2.4.2 3D point cloud generation, individual tree crown detection and crown metrics

An average of 209 images per plot were used to build 3D point clouds and derive orthomosaics. Images of insufficient quality (e.g. blurry images) were removed from the datasets. Each image was aligned and geo-referenced with the drone GPS logs using Agisoft Photoscan Professional 1.2.6 [41]. The drone-based GPS measurements provide higher accuracy than ground-based measurements under the dense canopies; we used more than 200 GPS points at each study plot from the geo-tagged images for georeferencing the whole plot map. The workflow included building dense point clouds, creating a mesh from the clouds, generating digital elevation models (DEM) and then generating the orthomosaics. 3D point clouds were generated using the Structure from Motion (SfM) technique [18,17] in Agisoft Photoscan Professional 1.2.6 software. An example of such an orthomosaic is depicted in Figure 2.

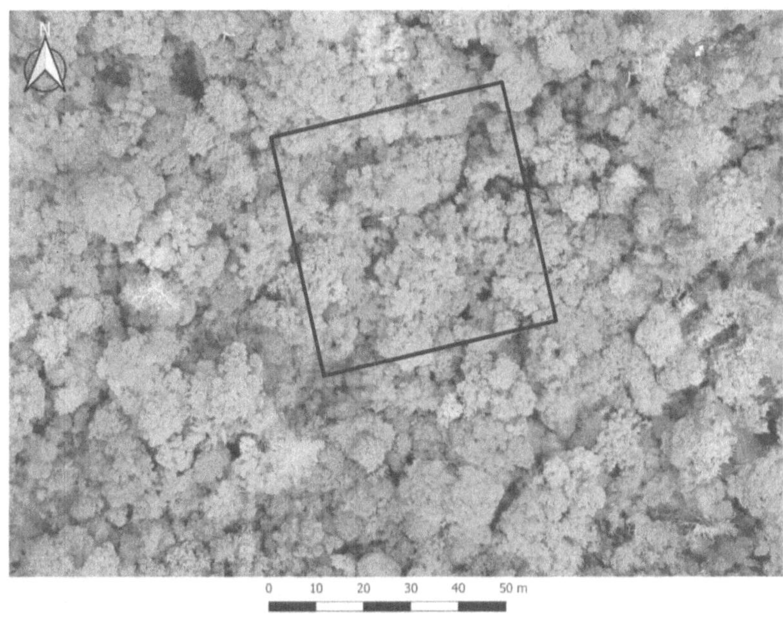

Figure 2. RGB orthomosaic image of one of the riparian study plots (HFr2).

Based on ground inventories, tree location information with tree ID and plot corner GPS coordinates were available. The tree location maps were in local Cartesian coordinates; they were georeferenced in UTM WGS 84 using the GPS points of the plot corners. The georeferenced location maps were overlain with the RGB orthomosaic images to manually identify the sap flux sample trees. The identification of the sample trees was based on the visible tree crowns with respect to the tree location points in the RGB orthomosaic image. To analyse the tree crown to water use relationship, the corresponding tree crowns were delineated manually through visual interpretations using QGIS 3.6 software [42] and cloud compare v.2.9 software [43]. Overall, the crowns of 42 out of the 72 sap flux sample trees could clearly be identified in the images, 5 in the upland plot and 3, 9, 13 and 12, respectively, in the four riparian plots. Due to this limited sample size for testing crown metrics vs. water use relationships, we pooled the data across all plots. The crown polygons were used for extraction of the point clouds with the lasclip function of the lidR R package [44]; for computing different crown metrics, the rLIDAR R package was used [45]. We extracted the metrics crown volume, crown projection area and crown surface area for the identified sap flux trees.

3.2.4.3 Automatic crown detection using AMS3D

In contrast to the manually delineated crowns of the 42 clearly identified sap flux trees, we followed an automated tree segmentation approach to detect and delineate the tree crowns of all other trees in the plots for scaling-up to stand transpiration; the time-consuming manual delineation for the hundreds of trees within a given stand would not be feasible. For all eight study plots, dense point clouds were extracted and the automated individual tree crown (ITC) detection algorithm AMS3D was applied [25]. AMS3D follows a non-parametric approach independent of pre-defined crown shape models and uses a multi-scale bandwidth technique that does not rely on single biophysical parameters. Due to its self-adaptive approach, which calibrates kernel bandwidth as a function of local tree allometric models, the segmentation process has the ability to characterize complex dense crowns and can deal with different crown shapes and multiple layers in the tropical forest [25]. As the AMS3D was previously only used with LiDAR, we adapted our high-resolution SfM point clouds to LiDAR standards by reducing the point density from 198 to 58 points m^{-2}. The point cloud density was reduced based on minimum distance between points as threshold criteria using the cloud compare v.2.9 software [43]. We cleared all points from the point clouds that lay below the minimum tree height of the respective plot to filter out non-canopy points and avoid interferences of single ground points in crown modelling. We then used the meanshiftR R package [46] which allows individual tree crown segmentation using the Adaptive Mean Shift 3D (AMS3D) clustering algorithm [25]. In our study, we calibrate the kernel bandwidth value based on the ratio of crown diameter and tree diameter as observed from ground inventories. In contrast, the original AMS3D approach uses local allometric models constructed from CHM to further calibrate kernel bandwidth. In our case, a CHM could not reliably be constructed from the point cloud data at our study plots due to the lack of clearly identifiable ground points. After ITC segmentation, we removed all crowns that comprised relatively low point cloud densities (fewer than 40 points per crown) in order to avoid irrelevant crowns (also see Aubry-Kientz et al. [31]). All individually segmented crowns of a given study plot were vectorised and crown metrics were computed analogously to the previously described methodology for manually selected trees.

We compared the number of segmented crowns per plot to ground stem counts (trees $\geq$ 10 cm DBH) and performed accuracy assessments by matching the ground-recorded stem locations of each tree to the centroid of delineated crowns. Matching was performed by finding the nearest neighbour distance within threshold criteria, i.e., a distance to nearest ground measured tree location below the segmented crown diameter of the tree (Figure A1). The threshold distance thus varied depending on the crown diameter. The accuracy assessment defines True Positives (TP), i.e., the detected trees match the actual trees in terms of tree location and threshold nearest neighbour distance, False Positives (FP) or commission error and False Negatives (FN) or omission error. From TP, FP and FN, the accuracy metrics precision (*Pr*), recall (*Re*) and F-score were calculated. *Re* indicates the tree detection rate, *Pr* indicates the correctness of the detected trees and the F-score is the overall accuracy considering both commission and omission errors [47].

3.2.5 Drone-based scaling, uncertainties and heterogeneity assessment of transpiration

To test the relationship between tree water use and different crown metrics, we used linear regressions, followed by residual plot analysis for normality and homoscedasticity tests. The allometric relationships from the linear regression served as the basis for scaling-up from individual tree water use to stand-level canopy E_t. The uncertainties associated with the scaling to stand E_t were compared among the different crown metrics and conventional ground-based approaches. Uncertainties in stand E_t estimates were assessed by bootstrapping the linear relationships between water use and the according predictor variables with the R package 'boot' (50,000 iterations) [48,49]. This yielded estimates of means for slope and intercept, as well as corresponding standard deviations as measures of uncertainty. For deriving stand E_t, the best performing crown metric (i.e. the metric with the lowest uncertainty) was applied to the stand-level crown datasets from the automated delineation algorithm. To test for differences in stand-level canopy E_t between upland and riparian plots, we performed an ANOVA. All statistical analyses were performed with R version 3.4.3 [50]. Plotting was performed using the Seaborn library [51].

3.3 Results

3.3.1 Tree water use vs. crown metrics

Out of the initial 72 sap flux sample trees 42 could clearly be identified and delineated in the aerial images and thus constituted the sample for further analyses. Their daily water use ranged from 8.5 to 95.7 kg day^{-1} (average of three sunny days). Crown volume and crown surface area ranged from 36 to 1604 m^3 and from 57.5 to 724.5 m^2, respectively.

Linear regression models between sap flux-derived daily tree water use and drone-derived crown metrics (Table A2) suggest highly significant linear relationships (P < 0.001) that explain 64% (crown volume) and 76% (crown surface area) of the observed variability in tree water use (Figure 3). For the conventionally applied ground-based inventory variable DBH, the regression model explained 38% of the observed variability (P < 0.001).

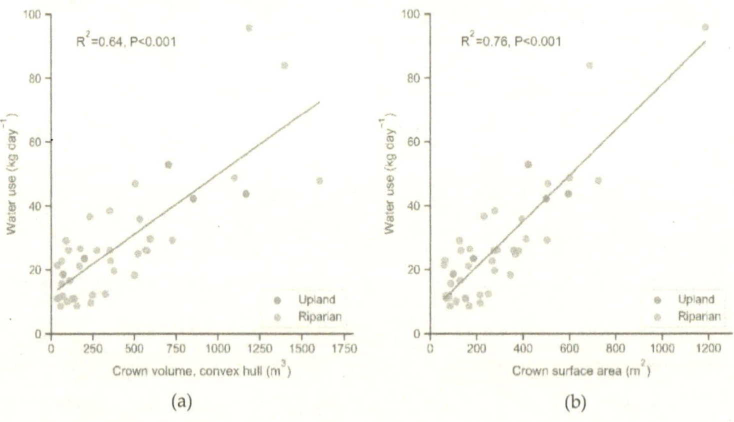

Figure 3. Tree water use vs. (a) crown volume and (b) crown surface area. Water use rates were estimated with sap flux techniques; 42 sap flux sample trees could be clearly identified in the aerial images (pooled data from upland and riparian plots). Crown metrics were derived from simultaneously carried out drone-based surveys.

3.3.2 Individual tree crown segmentation

The AMS3D algorithm produced between 140 and 181 segmented crowns per study plot. The difference between crown counts of the automated AMS3D approach and stem counts from the ground was 7% on average (Table 2). The F-score, which indicates the overall tree identification accuracy of the aerial method, had a moderate value of 60%, wherein a recall and precision of 65% and 56% were achieved, respectively (Table 2). The subsequently derived crown surface areas of the automatically segmented trees ranged from 7.7 to 1578.0 m^2, respectively (Table 3). A visualization of segmented crowns is shown as an example in Figure A2.

Table 2. Accuracy assessment of the automatically segmented trees by the AMS3D individual tree crown delineation method. Tree counts from the ground vs. aerial assessments, relative differences and accuracy metrics for each of the eight study plots.

	Ground-based counted trees	Drone-based detected trees	Difference (%)	True Positive	False Positive	False Negative	Precision	Recall	F-index
HF1	125	162	22.8	78	84	40	0.48	0.68	0.56
HF2	172	181	5.0	82	99	65	0.45	0.60	0.52
HF3	146	159	8.2	96	63	44	0.60	0.59	0.60
HF4	142	151	6.0	85	66	51	0.56	0.56	0.56
HFr1	135	140	3.6	74	66	31	0.53	0.68	0.60
HFr2	136	155	12.3	86	69	11	0.56	0.86	0.68
HFr3	157	140	12.1	101	39	33	0.72	0.54	0.62
HFr4	140	152	7.9	88	64	29	0.58	0.69	0.63
Mean	144	155	7.0	86	68	38	0.56	0.65	0.60

Table 3. Tree counts at the eight study plots based on automated crown segmentation using the AMS3D individual tree crown delineation method. Subsequently, the crown metrics volume, surface area and projection area were derived for all segmented trees.

Plot ID	Drone-based detected trees	Crown volume (m^3)			Crown surface area (m^2)			Crown projection area (m^2)		
		mean	min	max	mean	min	max	mean	min	max
HF1	162	409.1	5.4	2817.0	311.9	18.6	1217.0	30.5	3.3	165.0
HF2	181	338.9	3.9	3659.0	260.1	18.6	1292.0	35.6	2.5	249.6
HF3	159	512.1	1.7	4826.0	337.5	10.9	1578.0	39.3	2.2	299.5
HF4	151	610.3	1.4	4702.0	393.2	12.0	1493.0	45.9	2.5	251.9
HFr1	140	121.0	0.9	843.6	141.0	7.7	588.1	12.5	1.7	49.0
HFr2	155	61.6	0.7	427.6	90.8	7.8	353.2	10.0	2.3	38.4
HFr3	140	153.9	1.3	1068.0	155.9	10.1	562.7	21.4	2.3	126.6
HFr4	152	370.5	1.6	4221.0	275.7	9.9	1376.0	35.9	1.7	298.5

3.3.3 Canopy transpiration: scaling, uncertainties and spatial heterogeneity

The bootstrapping method suggests large differences in the uncertainties associated with the respective stand E_t estimates derived from crown metrics vs. conventional ground-based methods. As such, uncertainties when using crown surface area for E_t scaling were much smaller (17%) than when using the conventional DBH-based approach (51%) (Table 4). The drone-based E_t estimates ranged from 1.82 to 2.1 mm day^{-1} at the four upland plots and from 0.81 to 1.60 mm day^{-1} at the four riparian plots. Mean E_t was significantly higher (44%) in upland plots (1.9 ± 0.1 mm day^{-1}, mean ± SE) than in riparian plots (1.0 ± 0.2 mm day^{-1}, mean ± SE) (ANOVA, P = 0.004) (Table A3).

Table 4. Uncertainties associated with the scaling of transpiration from tree-level to stand-level based on different ground and drone-based methods. R^2 and P-values of linear regressions between plant water use and the according scaling variables. Uncertainties associated with scaling-up to stand transpiration based on these relationships, derived from parametric bootstrapping (with 50,000 iterations). N is the sample size of trees with sap flux measurements.

	Scaling approach	$\mathbf{R^2}$	**P value**	**Mean tree water use (WUmean)**	**Bootstrapped scaling uncertainty**
				kg day^{-1}	%
Ground-based measurements	N x WUmean	-	-	27.63 (measured)	67.5[1]
	DBH (cm)	0.38	<0.001	27.52	50.6
Drone-based crown metrics	Crown volume (convex hull, m^3)	0.63	<0.001	27.63	19.9
	Crown surface area (m^2)	0.76	<0.001	27.55	17.0
	Crown projection area (m^2)	0.68	<0.001	27.51	21.6

[1]no bootstrapping possible, instead CV-based approach Granier [52]

Based on the scaling variable with the lowest uncertainty (crown surface area, Table 4), we further assessed the spatial heterogeneity of E_t at different scales. Plot-to-plot heterogeneity of E_t was much higher among the four riparian plots (28.0% coefficient of variation, CV) than among the four upland plots (5.3% CV). In contrast, the relative within-plot variability of E_t was similar for riparian and upland plots (ANOVA, P = 0.72), with respective mean CV values of 30.1 % and 31.2%; however, the absolute within-plot variability of E_t was higher at the upland plots (Figure 4).

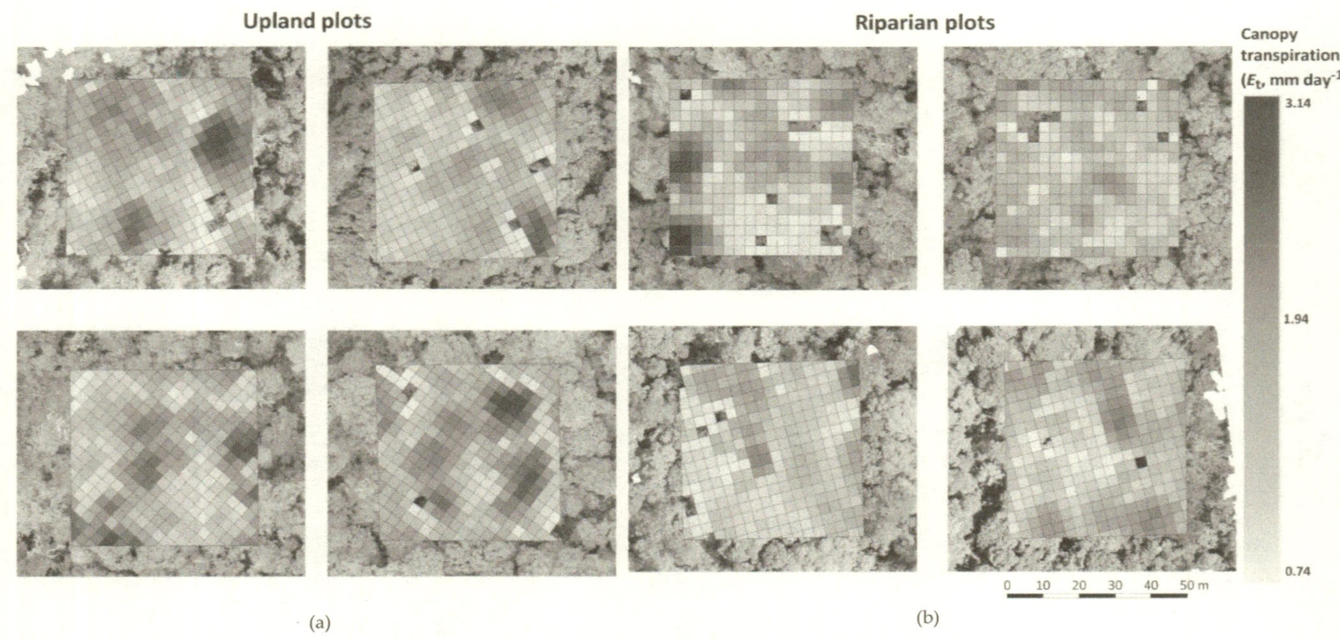

Figure 4. Spatial heterogeneity of canopy transpiration (Et, mm day^{-1}) within the (a) four upland and (b) four riparian study plots derived from the water use vs. crown surface area relationship (Fig. 3b). The blank tiles are due to the exclusion of low level canopies.

3.4 Discussion

Airborne tree crown detection in the studied tropical rainforest reduced the uncertainties in estimating canopy transpiration considerably. The newly established work flow resulted in scaling uncertainties from tree to stand of 17%, which is considered a great improvement compared to conventional DBH-based scaling (51% uncertainty). The predicted canopy transpiration suggests high stand-level differences between upland and riparian plots, with a 31% lower mean value at riparian plots, but higher plot-to-plot variation; these differences are driven by differences in crown packing among the plots. Likewise, the considerable variation of transpiration within plots is driven by local small-scale differences in crown packing. Overall, our study demonstrates the great potential of new drone-based methods for ecohydrological research, but it also points to some challenges.

Identifying the 72 sap flux sample trees in the aerial images proved to be difficult due to the dense and multi-layered canopy; only 42 of the sampled trees could be clearly identified to be used for further analyses. The 30 unidentifiable sap flux trees were uniformly distributed in terms of DBH. Therein, due to the linear relationship between crown surface area (or crown volume) and water use, all unidentified trees matter proportionally to their respective crown surface area (or volume). This seemingly stands in contrast to a previous study reporting over-proportional contributions of large emergent trees to stand E_t in old-growth tropical lowland forest [56]; however, this divergence is likely due to the lack of considerable emergent trees within our study plots in previously-logged lowland tropical forest. In previous studies applying airborne remote sensing approaches, the detection of small-statured trees was also reported to be particularly low and difficult [53,54]. For tree identification, we used tree location maps in local Cartesian coordinates drawn in ground surveys. These maps were georeferenced with the respective corner coordinates of the plots and subsequently overlain with their orthomosaics to locate the targeted trees from above. A clear identification was partially hindered by the lack of ground control points or tree markers, which would have likely facilitated the identification of smaller sub-canopy crowns within the dense forest canopy. Our attempts of letting helium balloons rise to the top of selected tree crowns (following [55]) were unsuccessful due to the high, dense and multi-layered canopies, wind and difficulty in controlling the balloons.

The high canopy closure of tropical forest canopies makes it difficult to classify ground points as a prerequisite for deriving CHMs from SfM point clouds. Thus, we opted for an exclusively point-cloud-based tree segmentation approach due to the reported enhanced performance in structurally diverse stands such as tropical forests [31]. We applied the self-adaptive approach called AMS3D, which calibrates kernel bandwidth as a function of local tree allometric models (Ferraz et al., 2016). Before applying this algorithm, we reduced the initial high density of our SfM point clouds (198 points m^{-2}) to the density (58 points m^{-2}) in order to increase the speed and quality of the clustering process [57]. Our adapted approach resulted in an overall moderate accuracy of tree detection (60%); however, the number of detected crowns at the plot-level was similar to ground stem counts, with a mean difference of less than 7%. Previous studies achieved higher detection rates, e.g., 69% in multi-layered Mediterranean forest [58] or 74% in French Guianian tropical rainforest [31], which is the best performance of an automated segmentation algorithm in a tropical forest so far. Other point cloud based methods such as Li2012 [27] performed well in woodlands dominated by few tree species, with accuracies over 81%; however, lower accuracy (<60%) was achieved in stands characterized by a dense canopy [59]. We further compared the automatically segmented crowns with the manually delineated crowns of the sap flux sample trees with respect to crown metrics. A linear regression model of automatically vs. manually derived crown surface area of the sap flux trees (forced through origin, $R^2=0.50$, $P<0.001$ has a slope of 1.25, suggesting that the automated algorithm on average overestimates the crown surface area of individual trees by 25% compared to manual delineation. However, due to the linear relationship between water use and the applied scaling variable, crown surface area, individual segmentation accuracy is not a constraint when assessing stand E_t: the sum of individual over- or under-segmented crowns within the plot boundaries will inevitably equal stand crown surface area and thus the predicted stand transpiration value.

Among the 42 identified sap-flux trees, we found close correlations between tree water use and crown metrics (best: crown surface area, $R^2 = 0.76$). Such a relationship has already been indicated for trees and palms in an agroforest [5]. Pooling these data suggests that a universal scaling may apply for trees but palms are different, and seem to follow another scaling factor (Figure A3). However, these relationships still need further exploration. In accordance with our results, several previous studies also explained variability in tree water use with crown or leaf area metrics [39,60–62]. Further studies from related ecological fields have also pointed to the high potential of using drone-derived crown metrics as predictors and scalars, e.g., for above ground biomass and canopy biomass [63].

Using crown surface area to scale-up from tree water use to stand-level E_t resulted in a substantial reduction of E_t estimation uncertainties compared to conventional ground-based approaches. In conventional approaches, DBH or DBH-derived sapwood area are used for scaling to E_t (e.g. [4,64], but associated uncertainties can be substantial when estimating E_t in heterogeneous stands [65]. Compared to the DBH approach (51% uncertainty), our crown-metric-based approach reduced scaling uncertainties (17%). This finding is in line with a previous study, where drone-derived crown volume substantially reduced E_t uncertainties in oil palm agroforests and monocultures [5].

The three upland plots with previous sap flux measurements were used for testing the quality of predictions. Although the sap flux was not measured concurrently, the results indicated low divergence of stand-scale means, with a much-reduced uncertainty (Table A3). Plotting stand E_t derived from crown metrics vs. E_t derived from conventional ground-based approaches for the eight study plots shows a significant linear relationship (R^2=0.56, $P<0.001$) and also suggests low divergence among the two methods (Figure. A4). The stand-level canopy E_t estimates derived from the new drone-based methodology were significantly higher for the four upland than the four riparian study plots. One may have expected that E_t at riparian sites is higher than at upland sites. However, a previous sap flux-based study at the same four upland plots showed no indications of soil water limitation of tree water use, in 2013 and 2014 (non-ENSO years) [39]. Further in accordance with our results, rubber and oil palm plantations in the lowlands of Sumatra had lower E_t at riparian sites than at upland sites [11]. Heterogeneity in E_t among sites at different topographic positions was also observed in other previous studies [6,7]. A study of Japanese cypress (*Chamaecyparis obtusa*) found E_t to be higher in valleys than at upland sites [66], while being similar for Japanese cedar (*Cryptomeria japonica*) [67]. In our study, the observed much lower E_t in riparian than in upland plots may be due to several factors. Rainforest species indicating disturbance (e.g. genus Macaranga) were more abundant in the riparian plots [68], and aboveground biomass was 43% lower than in upland plots [69]. The trees in the riparian plots were also smaller than in upland plots, which may go along with less turbulent energy exchange at the canopy level. Additionally, the position of the riparian study plots in moist landscape depressions probably induces higher air humidity at the canopy level and thus reduced atmospheric evaporative demand.

In addition to this spatial E_t variation between riparian sites and upland sites, we found varying plot-to-plot variation of E_t within each of the two categories. Relative E_t variation was low among the upland plots (5% CV) and more pronounced among the riparian plots (28% CV). These findings are in line with biomass assessments at these same sites that also point to a larger relative variability in the riparian sites than in upland sites [69]. Furthermore, our findings are in line with a previous study in an oil palm and rubber monoculture plantation in the same region, where E_t variability was enhanced by factors between 2.4 and 4.2 at (partially flooded) valley sites compared to adjacent upland sites [11].

We further used the new method to analyse spatial variability of transpiration within the plots. The canopy of the rainforest shows different degrees of crown packing, which we assessed with 3D canopy analyses; individually segmented trees would not be necessary but were used for calibration. The depicted differences in predicted transpiration per 9 m^2 tile of ground area (Figure 4) are driven by these local differences in crown packing. The minimum and maximum values of a derived 'crown surface area index' across the eight study plots were 0.18 and 32 m^2 m^{-2}, respectively. The strong differences in canopy packing result in the observed substantial small-scale variability of E_t; whether such small-scale differences can be considered realistic requires further investigation. Overall, our study underlines that topography and differences between riparian and upland forest sites exhibit E_t heterogeneity.

3.5 Conclusions

Crown surface area derived from drone-based imagery was a well-suited predictor of tree water use. In its application for scaling tree water use to stand-level transpiration, uncertainties were largely reduced compared to conventional diameter-based scaling approaches. The scaling was facilitated by an automated tree crown segmentation algorithm, which yielded moderately accurate results. Applying the method to the studied tropical rainforest in lowland Sumatra suggests large variations in spatial transpiration, both among and within study plots. Overall, we see great potential and improvement in drone-based methods for better understanding canopy structure and related ecohydrological responses in tropical forests and beyond.

Author Contributions: Conceptualization: J.A., A.R., D.H.; Formal analysis: J.A., A.R.; Funding acquisition: D.H.; Methodology: J.A., A.R., F.E.; Software: J.A.; Supervision: A.R., D.H.; Visualization: J.A.; Writing—original draft preparation, J.A.; Writing—review and editing: J.A., A.R., F.E., H., D.H.

Funding: This study was financially supported by the Deutsche Forschungsgemeinschaft (DFG) in the framework of a collaborative German–Indonesian research project (CRC 990 'EFForTS' project: sub-project A02). We further acknowledge support by the Open Access Publication Funds of Göttingen University and the DFG.

Acknowledgments: We would like to thank the Ministry of Research, Technology and Higher Education, Indonesia, for providing the research permit for field work (No. 285/SIP/FRP/E5/Dit.KI/VIII/2016 and No. 322/SIP/FRP/E5/Dit.KI/IX/2016). Furthermore, we would like to thank our field assistant Erwin Pranata for great support during the field campaigns. We would like to thank Katja Remboldt, Pierre Andre Waite, Fabian Brambach and Martyna Kotowska for providing the tree inventory data. Thanks to all 'EFForTS' colleagues and friends in Indonesia, Germany, and around the world.

Conflicts of Interest: The authors declare no conflict of interest.

CHAPTER 4

Multi-level temporal variation of sap flux densities in oil palm

Ahongshangbam J.[1], Röll A. [1], Hendrayanto [2], Hölscher D. [1,3]

[1] University of Goettingen, Tropical Silviculture and Forest Ecology, Germany

[2] Bogor Agricultural University, Forest Management, Indonesia

[3] University of Goettingen, Centre of Biodiversity and Sustainable Land Use, Germany

Correspondence to: Joyson Ahongshangbam, University of Goettingen, Tropical Silviculture and Forest Ecology, Buesgenweg 1, 37077 Goettingen, Germany. E-mail: jahongs@gwdg.de; Telephone: +49 (0) 551 39 12101; Fax: +49 (0)551 39 4019.

Manuscript in preparation (Advanced draft)

4.1 Introduction

Palms are diverse and abundant in both natural and in man-made ecosystems (Muscarella et al. 2020). Oil palm (*Elaeis guineensis* Jacq.) plantations have increased rapidly in recent decades, mostly in Southeast Asia (FAO, 2016). The transformation of tropical forest to monoculture plantations such as oil palm was reported to lead to biodiversity loss and substantial changes in productivity and biogeochemical cycles (Clough et al. 2016; Drescher et al. 2016; Röll et al. 2019). Severe changes in the hydrological cycle, e.g. increased periodical water scarcity and flooding, were also reported as a consequence of such land-use transformations (Merten et al. 2016, 2020). It is thus of general importance to better understand the water use characteristics of oil palms.

Studies on plant water use are often conducted with sap flux techniques. Niu et al. (2015) measured sap flux density in leaf petiole of oil palms using a heat dissipation sap flux method (Granier 1985) and presented calibrated, oil palm-specific parameters for the according sap flux equation. Measurements on stems of oil palms have, to our knowledge, not yet been performed. Therein, it would be of particular interest to examine the spatial variation of sap flux density across the radial and vertical profile of oil palm stems and to compare temporal dynamics at different levels, e.g. to sap flux patterns in leaf petiole. For trees, several previous studies reported strong radial sap flux gradients (Edwards and Booker 1984; Čermák et al. 1992; Phillips et al. 1996; Delzon et al. 2004; Link et al. 2020); assuming uniform sap flux density over the cross-section of the whole stem introduces substantial errors and uncertainties when estimating whole-tree water use (Čermák and Nadezhdina 1998; Ford et al. 2004; Kumagai et al. 2005). For oil palms, a scheme for scaling from sap flux point measurements in leaf petiole to whole plant water use and stand transpiration was proposed by Niu et al. (2015), but the option of scaling based on radial stem sap flux profiles is yet to be explored. Applying the leaf-based scheme, Röll et al. (2015) found that oil palm water use increases with increasing plantation age to approx. 8 years and then remains stable until the end of the rotation cycle (commonly 25 years). Oil palm water use of mature plantations was found to increase with increasing management intensity, and highly managed commercial plantation were found to surpass the transpiration rates of nearby forests (Röll et al., 2019). Hardanto et al., (2017) observed substantial heterogeneity in oil palm water use among partly flooded riparian sites and non-flooded upland sites. Even though many studies have added to a better understanding of the water use characteristics of oil palms in recent years, many of the underlying ecophysiological mechanisms remain unknown.

A previous study examined diurnal pattern of sap flux density in oil palms in relation to environmental drivers such as radiation and vapor pressure deficit and reported pronounced hysteresis, i.e. a decoupling of transpiration from environmental drivers. This was interpreted as a potential contribution of stem water storage to transpiration, resulting in early peaks of transpiration and a decline once the storage is depleted (Niu et al. 2015). Follow-up studies encompassing oil palm plantations of various ages (Röll et al. 2015) as well as simultaneous sap flux measurements and energy flux assessments with the eddy covariance technique (Meijide et al. 2017) also suggested a potential involvement of stem water storage mechanisms in oil palm transpiration. Important contributions to transpiration by internal water storage were previously described for some tropical tree species (Goldstein et al. 1998; Meinzer et al. 2004), subtropical trees (Oliva Carrasco et al. 2015) and temperate trees (Cermák et al. 2007; Köcher et al. 2013). For the arborescent palm species *Sabal palmetto* [(Walt,) Lodd, ex J, A, & J, H, Schult,], evidence of internal water storage was presented, and it was reported that the leaf water content was maintained by stem water storage up to 100 days when the soil water supply was depleted (Holbrook and Sinclair 1992). For oil palm, no studies focusing on stem water storage are available yet.

There are several ways of estimating the contribution of internal water storage to whole-tree water use. Asides from water balance approaches or cutting experiments, time lag analysis is one of the commonly applied approaches. Therein, sap flux density is measured at different vertical levels in the plant, usually along the stem; the timing of beginning transpiration in the morning, diurnal peak and transpirational decline (below a certain threshold) can subsequently be compared. Therein, the time lags allow calculating the amount of water that is removed from storage for transpiration (Pfautsch et al. 2015). Several previous studies compared time lags between canopy transpiration and sap flow in the base of the stem to estimates stem water storage (Goldstein et al. 1998; Phillips et al. 1999; Köcher et al. 2013). For woody species, it was reported that time lags are positively related to plant size (Oren et al. 1986; Goldstein et al. 1998; Phillips et al. 1999) and further depend on species-specific buffering capacitances (Edwards et al. 1986; Hunt and Nobel

1987) and on anatomical characteristics of the vascular system (Čermák et al. 1976). For oil palms, time lags along the stem have not yet been analyzed.

In our study, we conducted sap flux measurements in mature oil palms in lowland Sumatra, Indonesia. Sap flux density was simultaneously measured at different horizontal and vertical positions in the stem and near the base of leaf petiole. The main objectives of the study were 1) to assess the radial profile of sap flux density in the stems of oil palms, 2) to analyse hysteresis between sap flux densities and environmental drivers at different vertical levels, and 3) to evaluate the role of stem water storage for oil palm transpiration.

4.2 Materials and methods

4.2.1 Study area

The study was conducted in a commercial oil palm plantation (PTPN6, 1°41′35.0″S, 103°23′29.0″E) located in the lowlands of Jambi province, Sumatra, Indonesia. The terrain is flat with some small elevations; altitude is 76 m a.s.l. (Meijide et al, 2017). Annual mean precipitation and mean air temperature are 2235 mm yr^{-1} and 26.5°C, respectively (Drescher et al., 2016). The predominant soil type in the plantation are loam Acrisols (Guillaume et al. 2015). At the time of the field campaign in August 2018, the plantation was 16 years old; average palm height was 14.3 m with a stem density of 140 palms ha^{-1}. Leaf area index had been estimated to be 3.64 m^2 m^{-2} (Fan et al. 2015), with likely similar values at the time of our study.

4.2.3 Sap flux measurements

We conducted sap flux measurements in oil palms using three different sap flux methods: thermal dissipation probes (TDP, Granier 1985), the heat ratio method (HRM, Burgess et al. 2001) and the heat field deformation (HFD, Nadezhdina et al. 2012). TDP is based on a constant heating method (Granier 1985) and requires species-specific calibration (Lu et al. 2004); for oil palm petiole, this was performed by Niu et al. (2015), and new parameters for the sap flux equation were derived. The HRM is an improved modification of compensation heat pulse methods and, unlike the TDPs method, is capable of measuring very low and reverse fluxes (Burgess et al. 2001); when applied correctly, the method does not require species-specific calibration (Fuchs et al. 2017); however not tested for oil palm. HFD sensor use continuous heating and a combination of symmetrical and asymmetrical temperature measurements; they measure sap flux density at multiple depths into the xylem, thus allowing to derive radial sap flux profiles (Nadezhdina et al. 2012).

The measurement campaign was conducted from August to October 2018. Five mature oil palms were selected for the sap flux assessment. Their stem diameter at breast height (without bark and fronds) was 42.5 ± 3.3 cm (mean ± SE), their meristem height was 6.4 ± 0.3 m. TDPs were installed in four leaf petiole (approx. 30 cm from the base of the petiole, see Niu et al., 2015) of each oil palm. Two HRM sensors per measurement height were installed at the base (stem base) and the top of the stem (stem high). For the radial sap flux measurements we installed one HFD sensor per palm in the middle of the stem (stem mid) (at approx. 2.8 ± 0.4 m) Later during the experiment, two out of five oil palms were additionally equipped with HRM sensors in leaf petiole (four leaves per palm) and with HFD at the stem base and the top of the stem, always with enough circumferential distance to not interfere with the measurements of other sensors. A schematic diagram of a fully equipped oil palm is depicted in figure 1.

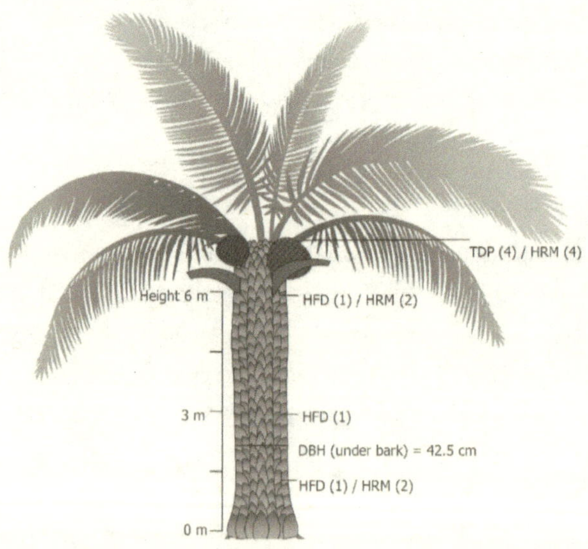

Figure 1: Multi-level sap flux measurement scheme on oil palms. Different types of sensors (TDP: thermal dissipation probe, HRM: heat ratio method, HFD: heat dissipation deformation) and their number per equipped palm are indicated below.

4.2.3 Data analysis and statistical methods

For TDP measurements, sap flux density J_s (g cm^{-2} h^{-1}) was calculated using the equation derived by Granier (1985), but with oil palm specific, calibrated equation parameters (Niu et al., 2015). Sap flux data from HRM and HFD sensors were processed using the software provided by the manufacturer (SapFlow Tool V. 1.4.1 ICT, Australia); thermal diffusivity as the only further input requirement for both sensors type was derived from the fresh weight and dry weight of an oil palm sapwood sample (n=15) (extracted from a depth of 1.5 cm into the stem). For all three sap flux methods, J_s are provided in 10 min intervals for further analysis.

For radial profiles of J_s as derived from mid-stem HFD measurements, data from four oil palms were used (one palm omitted due to sensor problems). Therein, the averaged values of J_s between 11:00 to 14:00 on a single sunny day were used. J_s data were normalized by setting the highest observed value for each palm to one, and the lowest to zero. Means and standard errors of J_s among the four palms were calculated for each measurement depth (0.5 to 7.5 cm, 1 cm steps).

To assess diurnal patterns of J_s at different vertical levels, J_s data recorded by HRM at the stem base, the top of the stem and in leaf petiole were normalized by setting the respective highest observed value to one, and the lowest to zero; data from a single sunny day was used for the analysis. Potential environmental drivers, i.e. vapor pressure deficit (VPD) and solar radiation (Rg) data, were also normalized. Normalized VPD and Rg were plotted against normalized J_s at different vertical levels to assess potential differences in drivers of J_s by examining hysteresis. Further, the respective areas of the hystereses were calculated.

Quantitative time lag analysis was conducted based on the timing of the onset (T_{onset}, peak (T_{peak}) and decline ($T_{decline}$) of the diurnal pattern of J_s. T_{onset}, T_{peak} and $T_{decline}$ were defined as the points in time where the normalized J_s data surpassed 0.1 in the morning, surpassed the near-maximum of 0.9 and fell back below 0.9 in the afternoon, respectively.

All statistical analyses and plotting were performed with R version 3.4.3 (R Development Core Team, 2016).

4.3 Results and discussion

Following the TDP scheme for oil palm leaf petiole by Niu et al. (2015), average daily palm water use was 194 ± 24 kg day^{-1} (mean ± SE, n=4 palms). This estimate agrees well with the overall high water use rates provided in previous studies in the same region, ranging from 158 to 249 kg day^{-1} (Ahongshangbam et al. 2019; Röll et al. 2019).

The multiple radial J_s measurements with HFD sensors in the mid-sections (stem mid) of four oil palms suggest that J_s is marginal in the outer stem (at 0.5 cm depth), reaches about 40% of its radial maximum at 1.5 cm depth and (near) maximum values at 2.5 depth. From 4.5 to 7.5 cm depth, values reach 30-60% of the maximum (Figure 2). Sap flux likely remains at non-zero levels deeper into the stem. To our knowledge, this is the first reporting of the radial sap flux profile of oil palm stems. For dicot trees, radial J_s patterns are typically highest at the outer edges of the stem and then gradually decline along the xylem radius (Phillips et al. 1996; Nadezhdina et al. 2002; Delzon et al. 2004; Granier et al. 1994; Wullschleger and King 2000; James et al. 2003). It was reported that this decline is stronger in larger trees compared to smaller trees (Čermák and Nadezhdina 1998). However, some studies on dicot trees also reported more variable radial sap flux patterns for certain tree species (Edwards and Booker 1984; James et al. 2002; Nadezhdina et al. 2002). In general, given the vast differences in the anatomy of monocot (oil) palms vs. dicot trees, e.g. in terms of xylem distribution, the contrasting radial J_s patterns do not come as a surprise. Our results show that J_s in oil palms was still substantial (~60% of its radial maximum) at a depth of 7.5 cm into the stem. It can thus well be that the whole cross section of oil palm stems is comprised of water conductive tissue (Killmann and Koh 1988).

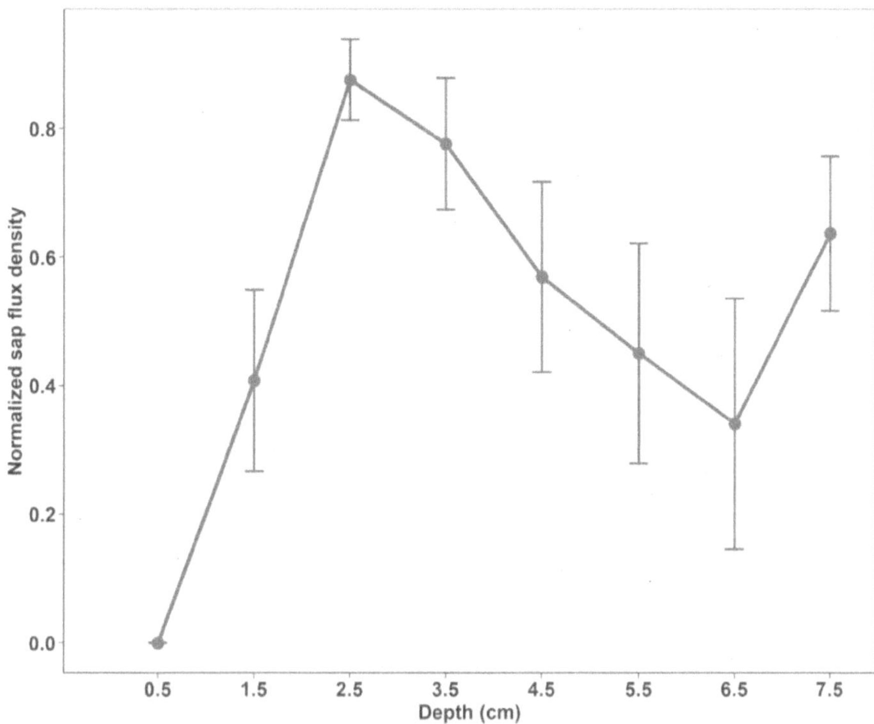

Figure 2. Radial sap flux density profile in the stem of oil palms, measured with HFD sensors at stem mid-level (2.8 ± 0.4 m). Normalization was performed by setting the respective highest observed value to one, and the lowest to zero. Points and bars represent means and standard errors of normalized J_s among four palms. Average stem radius (without bark) of the palms was 21.3 ± 1.7 cm.

Pronounced hysteresis was observed in the diurnal course of normalized J_s as measured with HRM vs. both VPD and Rg and at all three vertical levels, i.e. leaf petiole, stem high and stem base (Figure 3). Peak times of J_s generally coincided relatively closely with peak times of VPD and Rg (around 1:20 PM, refer table 1). Our results somewhat stand in contrast to previous studies that assessed the diurnal course of oil palm transpiration versus environmental drivers in the same study region; they reported an early peak of J_s (10 - 11 AM), i.e. before Rg and VPD; this resulted in large hysteresis of J_s, particularly to VPD (Niu et al. 2015, Röll et al. 2015). In our study, differences between VPD and Rg hysteresis were observed; where the areas within the hysteresis loop were larger for VPD than for Rg, indicating a closer coupling of J_s to Rg. This difference between VPD and Rg hysteresis varied substantially at the leaf-level (24% larger VPD hysteresis) and the stem (stem high (3%) and stem base (6.5%)). Also, the differences of the areas within the hysteresis loop were larger for VPD (26%) when compared between base stem and leaf level, but only 4% differences for Rg; suggesting VPD are more sensitive than Rg at the vertical levels of the oil palm. Large hysteresis in the water use response to environmental drivers are not scarce in existing literature; as such, earlier Rg peaks than J_s peaks have been described for several tree species (Zeppel et al. 2004; Dierick et al. 2010; Horna et al. 2011). For tropical bamboo species, the area of the hysteresis to VPD was 32% larger than for tropical trees while it was 50% smaller for Rg (Mei et al. 2016). Based on our hysteresis analysis, VPD are more sensitive than Rg and influenced more at the leaf level than stem of the oil palm.

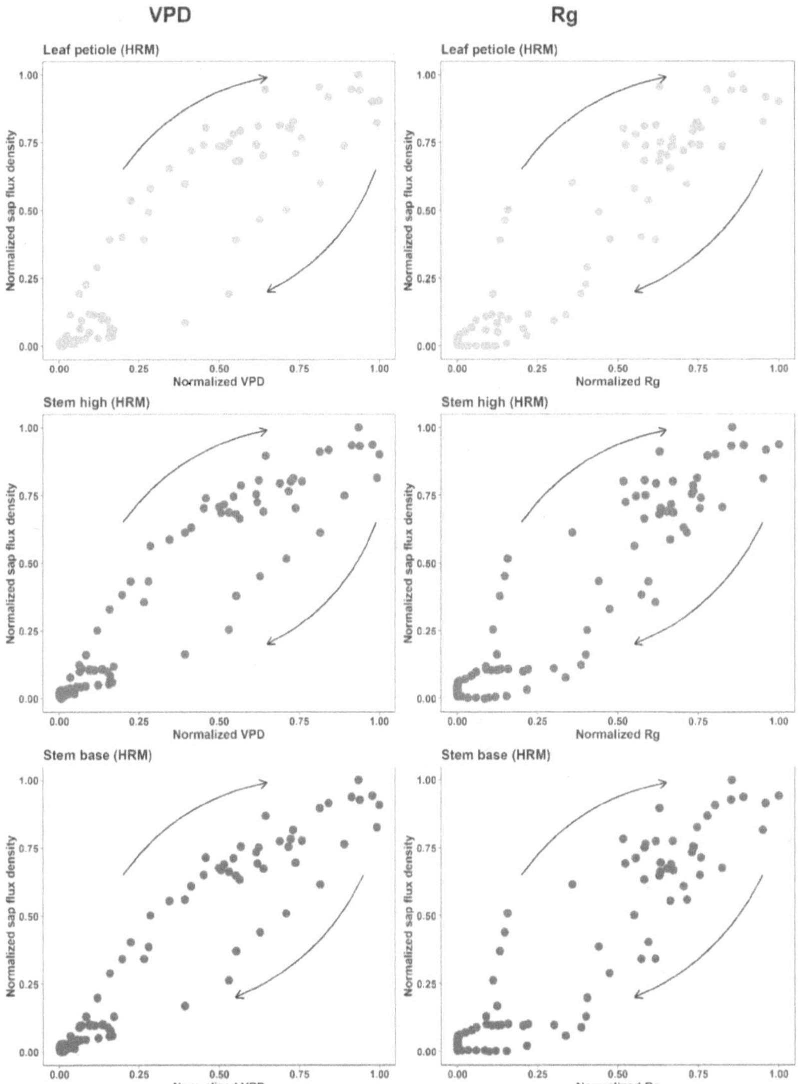

Figure 3. Normalized sap flux density at three different vertical level in oil palms (n=2 oil palms) plotted against (a) normalized VPD and (b) normalized Rg. 10 min averages on a single sunny day. Arrows indicate the direction in which the next observation in time occurred.

Time lag analysis of J_s at different vertical levels is a commonly applied tool to study the contribution of stem water storage to transpiration, the reasoning being that lags in onset, peak and decline of J_s represent water that is removed from or added to stem storage. In our study, we installed multiple sensor types at multiple vertical positions (stem base, stem mid, stem high, leaf petiole) to analyze the occurrence of time lags in the normalized J_s data of high temporal resolution (10 min steps). Generally, the diurnal course of J_s in oil palms followed the course of the environmental drivers VPD and Rg relatively closely (Table 1). All three applied sensor types (HRM, TDP, HFD) and vertical levels have a similar onset of J_s in the morning (7:40 to 8:20) (Figure 4, Table 1). The HRM measurements at the leaf-level show the earliest onset (7:40) while the TDP measurements on adjacent leaves show the latest onset (8:20am), indicating a possible methodological bias for the TDP measurements, which diverge from all other applied methods and levels. The same can be seen for the peak times: for all sensor types and vertical levels except the leaf-level TDP measurements (10:30), peak times start at 13:20 to 13:30 and thus correspond to the peaks of VPD and Rg. However, early peak of J_s measured by TDP were also observed in the previous studies at the same study site and same TDP sensors (Niu et al. 2015, Röll et al. 2015). Based on our results, care should be taken when using petiole-level TDP measurements for interpreting diurnal J_s patterns on (oil) palms, as they showed an early peak that two further methods did not detect; TDP-based estimates of daily transpiration rates, on the other hand, are only marginally affected by this methodological bias.

Near-zero J_s is reached at a similar time for leaf-level TDP and all levels of HRM (15:30 to 16:00), but it occurs much later (18:30 to 19:40) based on the HFD measurements (Figure 3, Table 1), which reached much deeper into the stem. These fluxes after sunset, which are not driven by VPD or Rg, potentially indicate refilling of stem water storage at night; both HRM and TDP leaf-level measurements indicate near-zero J_s at the time. The HFD recorded nighttime fluxes account for 9% of accumulated daily J_s. Considering only one method (HRM) and comparing normalized J_s in the stem base, the top of the stem and in leaf petiole, we found only small time lags and the diurnal curves largely overlapped (Appendix 1, Table 1). Our time lag analysis results thus leave only little room for a contribution of water storage to oil palm transpiration. They do not confirm previous speculations of strong contributions of stem water storage mechanisms to transpiration in oil palms (Niu et al., 2015; Röll et al. 2015). Other aspects of quantifying stored water in stem was to estimate the change in mass storage or balance difference between the base and top of the stem; however in our case absolute sap flow cannot be rely due to lack of calibration of the sap flux sensors, particularly measurements at the stem of the oil palm.

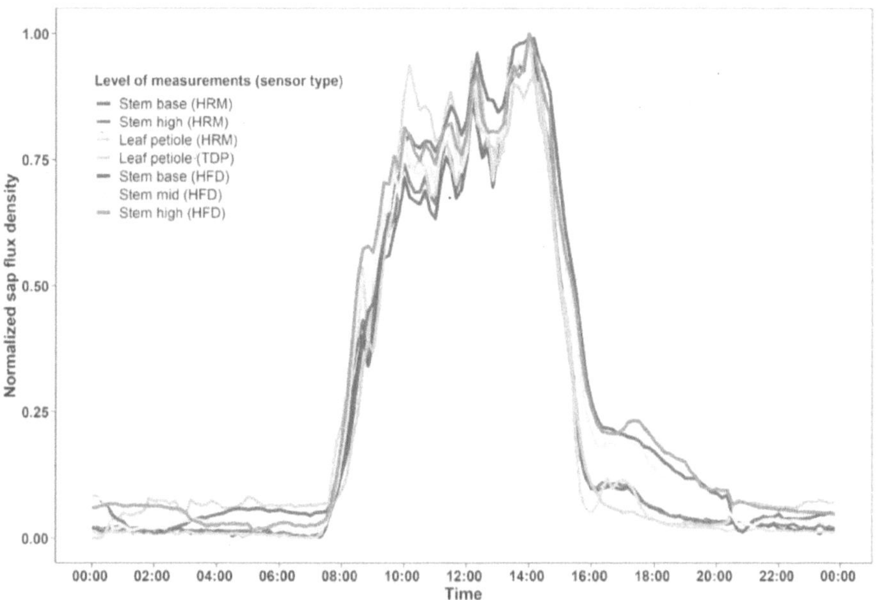

Figure 4. Diurnal pattern of normalized sap flux density in oil palms on a single sunny day (2nd-October-2018). Three different sap flux methods were applied at different vertical levels on the stem: heat ratio method (HRM) in stem base, top of the stem and in leaf petiole, thermal dissipation probes (TDP) in leaf petiole and heat field deformation sensors (HFD) at mid-stem.

For other palm species, previous studies reported strong contributions of stem water storage to transpiration. For the arborescent palm species *Sabal palmetto*, tree cutting experiments and water content measurements directly in the stem showed that 21-43% of the transpired water was withdrawn directly from the stem during imposed drought, and that the internal water storage had a significant role in maintaining leaf water content (Holbrook and Sinclair 1992). A study on the palm species *Washingtonia robusta* (H. Wendl.) found 28 min (8 m tall palm) to 44 min time lags (28 m tall palm) between boles and petiole based on TDP measurements (not calibrated); this was interpreted as a withdrawal of water from stem storage, which increases with stem size (Renninger et al. 2009). These studies stand in contrast to our results, which may indicate that the role and functioning of internal stem water storage may also differ substantially among different palm species. Substantial time lags between stem sap flow and canopy transpiration were further reported for several temperate tree species (Schulze et al. 1985; Zweifel and Häsler 2001; Cermák et al. 2007; Köcher et al. 2013) and savanna trees (Scholz et al. 2008). For tropical forest trees, Goldstein et al. (1998) observed large time lags in large individuals (4 to 5 hours) but concurrent sap flux in smaller trees. In contrast, time lags smaller than 20 minutes were observed between the stem and top branches of 50 m tall emergent trees in Bornean tropical forest (Kume et al. 2008). Likewise, Chen et al., (2016) reported no time lags between basal and crown sap flux in tropical lianas. Looking at these partially contradictory results, it should be kept in mind that time lags in sap flux at different vertical levels are a mere indication of a potential contribution of stem water storage to transpiration. Therein, the lack of observed time lags does not necessarily indicate the absence of stem water storage. Time lags, or the lack thereof, may be influenced by several factors such as axial hydraulic resistance, wood density, environmental controls, stomatal openings or tree size (Holbrook, 1995; Cermák et al. 2007). In our study, we add to this by providing first insights on time lags among multiple vertical levels in oil palms. The time lag analysis suggests that stem water storage does not have a substantial contribution to transpiration in oil palms. However, other stem water storage mechanisms that do not translate into time lags at different vertical positions may be at play, which makes an interesting subject for future inquiry.

Table 1: Temporal variation of sap flux densities (J_s) at different vertical levels in stem and leaf petiole of oil palm. Data from one single sunny day (see Fig. 3); all data were normalized. T_{onset} is the time it takes for normalized J_s to rise from 0.1 to 0.9 of the maximum, T_{peak} is the time when J_s is over 0.9 and $T_{decline}$ is the period it takes for J_s to fall from 0.9 to 0.1. Start and end times and durations of the respective periods are displayed for all vertical levels and sensor types.

	T_{onset}		T_{peak}		$T_{decline}$	
	Start-end time (hh:mm)	Duration (hh:mm)	Start-end time (hh:mm)	Duration (hh:mm)	Start-end time (hh:mm)	Duration (hh:mm)
VPD	8:10 to 13:30	5:20	13:40 to14:30	0:50	14:40 to 18:20	3:00
Rg	7:10 to 13:20	6:10	13:30 to 14:10	0:40	14:20 to 17:00	2:40
Stem base (HFD)	8:00 to 13:10	5:10	13:20 to14:30	1:10	14:40 to 19:40	5:00
Stem mid (HFD)	7:50 to 13:10	5:20	13:20 to 14:20	1:00	14:30 to 18:30	4:00
Stem high (HFD)	7:50 to 13:20	5:30	13:30 to 14:20	0:50	14:30 to 19:30	5:00
Stem base (HRM)	8:00 to 13:20	5:20	13:30 to 14:20	0:50	14:30 to 15:50	1:20
Stem high (HRM)	7:50 to 13:10	5:20	13:20 to 14:20	1:00	14:30 to 16:00	1:30
Leaf petiole (HRM)	7:40 to 13:10	5:30	13:20 to 14:00	0:40	14:10 to 15:30	1:20
Leaf petiole (TDP)	8:20 to 10:20	4:00	10:30 to 14:10	2:40	14:20 to 16:00	1:40

4.4 Conclusion

To our knowledge, this study provides first insights on horizontal and vertical patterns of sap flux densities in oil palms. The radial profile shows lower at the outer margin, increased to 2.5 cm under the bark and remained high to the innermost measured depth at 7.5 cm. The temporal dynamics of sap flux densities at different vertical levels in the stem and in petiole show relatively small time lags. Based on time lag analysis there is thus little room for a contribution of stem water storage to oil palm transpiration. However, other stem water storage mechanisms that do not translate into time lags at different vertical positions may be at play, which may be the subject of future inquiry.

CHAPTER 5

SYNTHESIS AND OUTLOOK

5.1 Overview

This book provides more insights about the tree and oil palm water use in terms of scaling scheme, spatial heterogeneity and multi-level temporal dynamics in the lowland of Sumatra, Indonesia. We conducted sap flux measurements in oil palm monoculture, oil palm agroforest and tropical rainforests. Simultaneously, drone imageries were taken to create the 3D structure of the stands and further, delineate the individual tree or palm crown metrics. We tested the different crown metrics to predict the stand transpiration (E_t) in oil palm, agroforest trees and tropical trees and associated uncertainties were estimated (Chapter 2 & 3). We further assessed the heterogeneity of E_t between oil palm agroforest and oil palm monoculture (Chapter 3). In a tropical forest, heterogeneity of E_t between/among upland and riparian sites was assessed based on drone-based crown metrics with the addition of automatic tree crown detection approach (Chapter 3). We addressed the radial pattern of sap flux density of oil palm and multi-level temporal dynamics of oil palm water use to understand the role of stem water storage in oil palm water use (Chapter 4).

5.2 Scaling variable and its associated uncertainties

We tested different crown metrics to predict tree and palm water use. For oil palm, crown volume convex hull explained 69% ($P = 0.01$) of the observed palm-to-palm variability in daily water use For agroforest trees, crown volume models with an alpha level 0.25 explained 81% ($P < 0.001$) of tree-to-tree variability across the studied species. While, for tropical trees, the crown surface area explained 76% ($P < 0.001$) of the observed variability in tree water use. There was no single linear crown volume model that fit both oil palms and agroforest trees; however, combining these data with tropical trees suggests that a universal scaling may apply for trees but palms follow different scaling factor (Figure 5.1). But, these relationships still need further exploration.

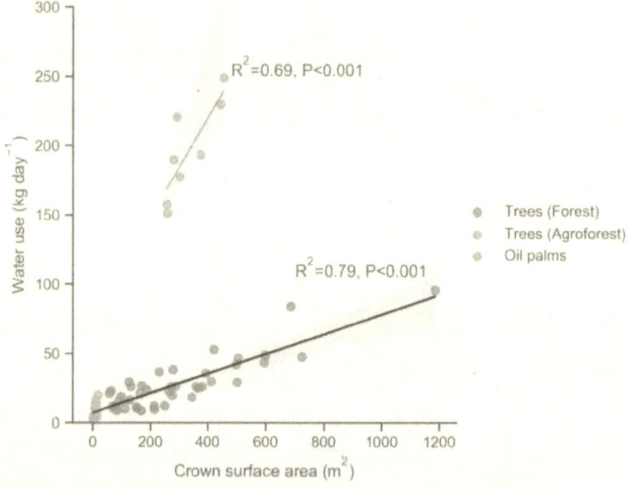

Figure 5.1 Relationship between water use and crown surface area for all the pooled datasets (oil palms, agroforest trees and tropical forest trees).

For the conventional DBH, the linear regression model explained 65% ($P < 0.001$) and 38% ($P < 0.001$) of the observed variability of the agroforest trees and tropical trees respectively. While for oil palms, no significant ground-based explanatory variables were available for comparison. In several studies, DBH yielded an R^2 of 0.65, which is quite similar to our agroforest trees and also the relationship between DBH and tree water use was found to hold across species (Granier et al., 2000; Schiller et al., 2007; Yue et al., 2008). Whereas in a premontane forest in Costa Rica, the correlation of water use to DBH showed differences among species (Moore et al., 2017). Likewise, species-specific trajectories were suggested by reforestation in the Philippines (Dierick & Hölscher, 2009). There are further general concerns in using the diameter for scaling.

The bootstrapping method suggests large differences in the uncertainties associated with crown metrics vs. conventional ground-based methods while scaling up the stand transpiration (E_t). For agroforest trees, the crown volume is associated with an uncertainty of 28% whereas uncertainty of 100% is associated when using DBH for scaling. For the oil palms in the agroforest and the monoculture, the uncertainty estimates associated with crown volume scaling were 37% and 35%, respectively. In the case of tropical trees, uncertainties when using the crown surface area for E_t scaling were much smaller (17%) than when using the conventional DBH-based approach (51%). From our results, we understand that the crown metrics are suitable scaling variable and reduced uncertainties largely for stand transpiration estimations. Several other studies also explained variability in tree water use with crown or leaf area metrics (Hatton & Wu, 1995; Oren et al., 1999; Röll et al., 2019). Also, our study showcases the applicability of low-cost drone imageries in the deriving 3D structure of the forest and delineating individual tree crown metrics; which will support the forest inventory in the future. Further studies from related ecological fields have also pointed to the high potential of using drone-derived crown metrics as predictors and scalars, e.g., for above-ground biomass and canopy biomass (Vauhkonen et al., 2014). We thus found that the crown metrics derived from drone-based photogrammetry are suitable variable for scaling stand transpirations.

5.3 Spatial heterogeneity of tree and oil palm water use

We applied the drone-based approach to estimates E_t in the tropical rainforest where it is more complex, heterogeneous and undulating terrain site conditions (upland and riparian) and assessed the spatial heterogeneity of E_t between upland and riparian sites. The stand-level canopy E_t estimates were significantly higher (44%, ANOVA, P = 0.004) for the four upland (1.9 ± 0.1 mm day^{-1}, mean ± SE) than the four riparian study plots (1.0 ± 0.2 mm day^{-1}, mean ± SE). Our result observed much lower E_t in riparian than in upland plots; which may be due to several factors such as more disturbance (e.g. genus Macaranga) (Rembold et al., unpublished), and lower aboveground biomass (43%) in the riparian plots lower than in upland plots (Kotowska et al., unpublished). Rubber and oil palm plantations in the lowlands of Sumatra had lower E_t at riparian sites than at upland sites (Hardanto et al., 2017); which is in agreement with our results. Other studies also observed the spatial heterogeneity of E_t among sites at different topographic positions (Kume et al., 2016; Loranty et al., 2008; Mackay et al., 2018)

We further assessed the spatial heterogeneity of E_t at plot-to-plot scale and within-plot scale. Plot-to-plot heterogeneity of E_t was much higher among the four riparian plots (28.0% coefficient of variation, CV) than among the four upland plots (5.3% CV). This is in agreement with biomass assessments and a variability at these sites (Kotowska et al., unpublished). Also, in oil palm and rubber plantation in the same region, E_t variability was enhanced by factors between 2.4 and 4.2 at (partially flooded) valley sites compared to adjacent upland sites (Hardanto et al., 2017), which is similar with our results.

Assessing the relative within-plot variability, E_t was similar for riparian and upland plots (ANOVA, $P = 0.72$), with respective mean CV values of 30.1 % and 31.2%; however, the absolute within-plot variability of E_t was higher at the upland plots. The observed heterogeneity within the plot is mainly due to the local differences in crown packing. Such small-scale variability of E_t would require further assessments related to crown structure and packing.

On the other hand, we also assessed the differences of transpiration between oil palm monoculture and oil palm agroforest (EFForTS-BEE) at the individual and stand level. At the individual level, the daily water use per palm was higher (32%) in the agroforest than in the monoculture (ANOVA, $P < 0.01$), ranged between 158 and 249 kg day^{-1}; which is mainly due to the thinning of oil palm in agroforest while setting up the experimental site (Teuscher et al., 2016) and thus provides less competition for light, soil water and nutrients for the remaining oil palms. Previously, 36% higher per-palm fruit yield in thinned agroforests than in untreated monocultures was reported (Gérard et al., 2017), which agrees well with our result. While, the studied monoculture is relatively intensively managed, with fertilizer application including 230 kg N ha^{-1} year^{-1} (Teuscher et al., 2016). Comparing with small-holder oil palm plantation of similar age (108 ± 8 kg day^{-1}, mean ± SE among eight sites), the observed water use rates of oil palms was higher and compare well with commercial oil palm monoculture plantation (intensively managed) in the region (178 ± 5 kg day^{-1}) (Meijide et al., 2018; Röll et al., 2015). Thus, our data indicate that management intensity leads to variability of oil palm water use rates.

At the stand level, the stand transpiration (E_t) of the oil palm agroforest (1.9 mm day^{-1}) was 37% lower than in the oil palm monoculture (3.0 mm day^{-1}), which is contrast at individual palm level. The higher per-palm water use in the oil palm agroforest thus did not compensate for the reduction in oil palm stand density when scaled to the stand level. The 3-year old, comparably small inter-planted trees in the agroforestry plot contributed rather little to overall stand transpiration (15%). Future studies, particularly with the full grown trees, would definitely provide more detail information about the water use pattern of such an oil palm agroforestry experiment and thus, would support in reducing the impact of oil palm cultivation on biodiversity and ecosystem functioning.

5.4 Radial flux and temporal variations of oil palm water use

Our study provides the first data of the radial sap flux profile in the stem of oil palm. We found that the sap flux density was lower at the outer part and peaked at 2.5 cm depth and remained high until the end of the sensor depth (7.5 cm), which is in contrast with dicot trees where higher sap flow was observed at the outer edges of the stem and gradually decline towards the centre (Delzon et al., 2004; Nadezhdina et al., 2002; Phillips et al., 1996). Commonly, the sap flux decline gradually along the entire xylem radius, as reported on other studies on trees (softwood or hardwood) (Granier et al., 1994; James et al., 2003; Wullschleger & King, 2000) or varies across the sapwood area (Edwards & Booker, 1984; James et al., 2002; Nadezhdina et al., 2002). While estimating whole-tree water use, assuming uniform sap flux across the radial direction leads to high errors and uncertainties (Čermák & Nadezhdina, 1998; Ford et al., 2004; Kumagai et al., 2005). In the case of oil palm, estimation of individual oil palm water use was setup and calibrated based on leaf-level measurements in previous studies (Niu et al., 2015); the possibility of the stem with radial sap flux profile consideration may benefit the accurate estimation of palm water use. Also, it would be interesting to understand more about the stem anatomical structure of oil palm and its relationship to radial water use patterns.

We assessed the influence of environmental drivers against the diurnal course of sap flux density (J_s) and compared between different vertical levels of the oil palm. A pronounced hysteresis was against both VPD and Rg but different hysteresis of J_s against VPD and Rg was observed, more clearly at leaf level with an area of hysteresis difference of 24%. While comparing the hysteresis of J_s at stem base and leaf level, we found that large difference in the area of hysteresis against VPD (26%) and less difference against Rg; indicating that VPD was more sensitive at the vertical levels of the oil palm. Previous studies also reported a pronounced hysteresis but with an early peak of J_s (10 - 11 AM) before Rg and VPD (Niu et al. 2015, Röll et al. 2015), which is in contrast with our observations. In the case of tropical bamboo species, the area of the hysteresis to VPD was 32% larger in bamboos than in trees while 50% smaller against Rg (Mei et al., 2016). Large hysteresis in the water use response to environmental drivers was observed in many dicot tree species; where the Rg peaks first than J_s (Dierick et al., 2010; Horna et al., 2011; Zeppel et al., 2004). Such hysteresis may link to other mechanisms such as stomatal sensitivity, hydraulic conductance. Internal stem water storage of the stem of oil palm may also play a certain role in oil palm water use but no information available for oil palm.

We further assessed the time lag analysis of the diurnal patterns of J_s at the different vertical levels of the oil palm. We found no or little time lag differences in the sap flux pattern between the stem base, stem mid, stem high and leaf petiole in oil palm. Our results suggested that the small observed differences based on different timing of sap fluxes; thus, would leave little space for a contribution of water storage to transpiration and do not confirm the previous speculations of strong contributions of stem water storage mechanisms to transpiration in oil palms (Niu et al., 2015; Röll et al. 2015; Meijide et al. 2017). Commonly, mass balance between the base and top of the trees was also analysed along time lag analysis for stem water storage estimations; but in our case, absolute sap flow cannot be used due to lack of sensors calibration; thus require further investigations for detailed understanding of the role of stem water storage in the oil palm. In case of other palm species, a significant role of internal water storage was reported e.g. in arborescent palm *Sabal palmetto*, the transpiration loss was directly withdrawn (21 to 43% of the total loss) from the stored stem water during imposed drought (Holbrook & Sinclair, 1992) while in palm (*Washingtonia robusta*), the time lag difference between petiole-bole was 44 min and 28 min respectively in 28 m tall palm and 8 m tall palm (Renninger et al., 2009). Our results in comparison with these palms indicate that the role and functioning of the stem internal water storage may differ among palms. Moreover, palms are diverse and they are abundant in natural and man-made ecosystems (Muscarella et al., 2020). In the dicot tree, larger time lags between the stem and leaf-level transpiration were reported (Cermák et al., 2007; Goldstein et al., 1998; Scholz et al., 2008; Schulze et al., 1985; Zweifel & Häsler, 2001). Our results thus provide the first sights information about the stem water storage in oil palm based on time lag analysis but further studies incorporating other factors such as stomatal conductance, stem anatomical structure, hydraulic pathway and tree size would provide detailed information on oil palm stem water storage studies.

In conclusion, our study highlights that the crown metrics derived from drone-based imagery predicted tree and palm water use quite well. Such a scaling variable at the whole-plant level was previously not available in the case of oil palms. Associated uncertainties while scaling up also reduced largely as compared to conventional DBH approaches. Large differences in individual palm water use and stand transpiration between oil palm agroforest and oil palm monoculture clearly witnessed. In a tropical rainforest, spatial heterogeneity of stand transpiration exhibit between upland and riparian sites and also both among and within study plots. Radial sap flux pattern at the stem of oil palm was encouraging and firstly reported in our study. Diurnal patterns of oil palm water use were influenced by environmental drivers, more at the leaf level and time lag differences suggested that there is the little role of stem water storage in oil palm water use; however there might be other underlying mechanisms which may be subject to future investigations.

5.5 Future scope

We explored the potential and applicability of drone imageries (RGB) in construction of the forest 3D structure using photogrammetric techniques and uses in scaling plant water use from individual plants to the stand level. Additionally, the automatic tree segmentation in the 3D structure provides a new aspect of tree crown delineation in the tropical forest. Overall, we see great potential and improvement in drone-based methods for a better understanding of canopy structure and related ecohydrological studies in tropical forests and beyond.

Furthermore, a drone equipped with thermal imageries provides the canopy or land surface temperature data of the vegetation or forest. Based on surface energy models, such a drone-based approach showed recently a great potential in estimating the evapotranspiration of the relatively larger area in oil palm plantation (Ellsäßer et al., 2020). Developing and optimizing such methods would support measuring the ecohydrological responses of the forest or vegetation at larger scales in the future.

The Ecosystem Spaceborne Thermal Radiometer Experiment on Space Station (ECOSTRESS, https://ecostress.jpl.nasa.gov) is one of the aspects where we can look into it for the future ecohydrological studies at a larger scale. It provides a range of datasets ranging from LST, Evapotranspiration (ET), Evaporative stress index and water use efficiency at a spatial resolution of 70 m x 70 m. This newly launched data would ultimately help in ecohydrological studies; particularly in the lowland of Sumatra where large land-use transformations were evident. However, available of datasets in this area is limited due to cloud cover throughout the year. Here, one sample example of the ET map in Indonesia is provided (Figure 5.2) from the Evapotranspiration PT-JPL model (Fisher et al., 2020); one of the ECOSTRESS products (Hook, S. & Fisher, J., 2019). These dataset has great potential to estimate or scale-up large scale ET or understand the spatial heterogeneity on the different land-use type (as shown in Figure 5.3) and can assess temporal dynamics of ET, may be seasonal, not diurnal. On the other hand, LST product (Hook, S. & Hulley, G., 2019) can be input and optimized other independent surface energy model to estimate ET for large scale, as similar in (Ellsäßer et al., 2020) but with this satellite data (if available).

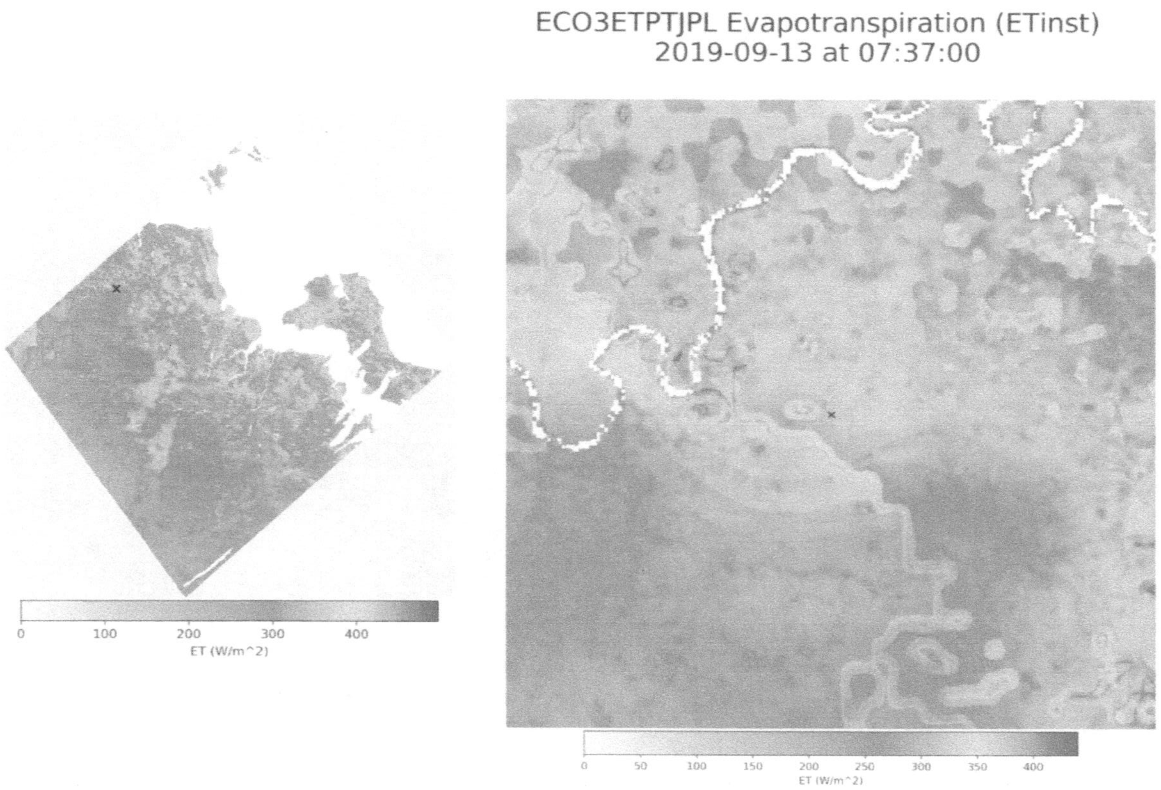

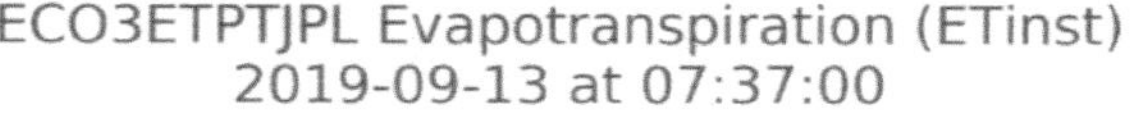

Figure 5.2 Instantaneous Evapotranspiration map of Sumatra, measured by ECOSTRESS PT-JPL datasets on 13 September 2019 at 7.37 AM local time. The black cross (X) in the map denotes the location of PTPN6 eddy covariance tower, measured an ET value of 315 W m^{-2} at this particular time. The right side shows the zoomed picture of the tower site in the oil palm plantation.

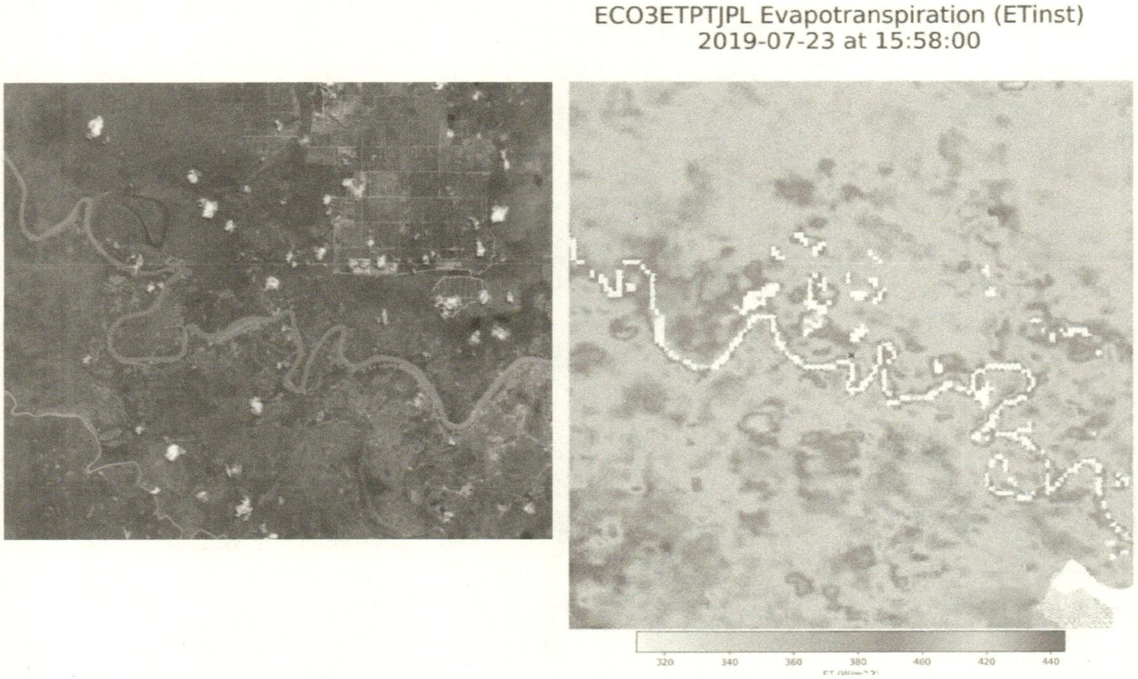

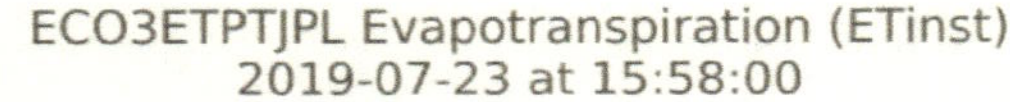

Figure 5.3 RGB images from Google Earth (left) and Instantaneous Evapotranspiration map (right) of (near Palembang) Sumatra, measured by ECOSTRESS PT-JPL datasets on 23 July 2019 at 3.58 PM local time. The right side shows the variability of ET in different land use patterns such as oil palm plantation, oil palm barren land, settlement.

www.ingramcontent.com/pod-product-compliance
Lightning Source LLC
LaVergne TN
LVHW040911150826
845672LV00007B/1985

* 9 7 8 3 3 8 4 2 2 3 4 2 5 *